Woodrow Wilson
International
Center
for Scholars

REACHING ACROSS THE WATER

International Cooperation
Promoting Sustainable
River Basin Governance
in China

JENNIFER TURNER & KENJI OTSUKA

May 2006

Woodrow Wilson International Center for Scholars

Available from the China Environment Forum
Woodrow Wilson International Center for Scholars
One Woodrow Wilson Plaza
1300 Pennsylvania Avenue, NW
Washington, DC 20004-3027

www.wilsoncenter.org/cef

Research Assistants: Linden Ellis, Timothy Hildebrandt,
Charlotte MacAusland, Louise Yeung, and Lulu Zhang
Chinese Translator: Serena Yi-ying Lin
Japanese Translator: Kenji Otsuka
Graphic Design: Lianne Hepler

ISBN 1-933549-06-8

Cover photo: The valley above Tiger Leaping Gorge, where
a major dam is being planned. (Photo Credit Ma Jun)

The Woodrow Wilson International Center for Scholars

The Woodrow Wilson International Center for Scholars aims to unite the world of ideas to the world of policy by supporting preeminent scholarship and linking that scholarship to issues of concern to officials in Washington. Congress established this non-partisan center in 1968 as the official, national memorial to President Wilson. The Center has 13 projects and programs, sponsors an international fellows program, and has an independent Board of Trustees (comprised of ten citizens, appointed by the President of the United States, and nine government officials—including the Secretary of State). The Center's president is the Honorable Lee H. Hamilton and Joseph B. Gildenhorn chairs the Board of Directors. www.wilsoncenter.org

The China Environment Forum

Since 1997, the Woodrow Wilson Center's China Environment Forum (CEF) has sought to initiate sustainable development approaches in China by: promoting information sharing, facilitating policy debates, and, most importantly, building networks between U.S., Chinese, and other Asian policymakers, nongovernmental organizations, researchers, businesses, and journalists to resolve common environmental and energy problems. CEF regularly brings together experts with diverse backgrounds and affiliations—including U.S. and international specialists from the fields of energy, environment, China studies, economics, and rural development. Through monthly meetings and the yearly journal *China Environment Series*, CEF aims to identify the most important environmental and sustainable development issues in China and explore creative ideas and opportunities for governmental and nongovernmental cooperation. www.wilsoncenter.org/cef

Institute of Developing Economies

The Institute of Developing Economies (IDE) was founded in 1960 as a statutory organization under the jurisdiction of the Ministry of International Trade and Industry (now the Ministry of Economy, Trade and Industry), acting as a social science institute of basic and comprehensive research in the areas of economics, politics and social issues in developing countries and regions. Since the merger with the Japan External Trade Organization (JETRO) in July 1998, the institute aims to promote the expansion of trade relations and economic cooperation with all developing countries and regions, including Asia, Middle East, Africa, Latin America, Oceania, and East Europe. JETRO was reorganized into an Incorporated Administrative Agency on 1 October 2005. www.ide.go.jp

ABOUT THE AUTHORS

Jennifer L. Turner has been the coordinator of the China Environment Forum at the Woodrow Wilson International Center for Scholars since 1999. As coordinator she organizes meetings and study tours, and creates publications that aim to promote dialogue and collaboration between U.S., Chinese, and other Asian government, NGO, research, and business communities to help solve environmental problems in China. Her current research focuses on water policy and the growth of "green" civil society in China.

Kenji Otsuka is a researcher in the Environment and Natural Resource Studies Group within the Inter-disciplinary Studies Center of the Institute of Developing Economies (Japan External Trade Organization). He has conducted joint research projects with Chinese scholars and researchers since 1993 at the institute. His fieldwork with Chinese counterparts has covered environmental policy process, water resource and river basin governance, environmental pollution disputes, environmental awareness and community/nongovernmental organization activism in China.

TABLE OF CONTENTS

PREFACE AND
ACKNOWLEDGEMENTS

T his report on water challenges facing China developed out of a joint project be-
tween Jennifer Turner at the Woodrow Wilson Center's China Environment
Forum (WWC/CEF) and Kenji Otsuka at the Institute of Developing Economies,
Japan External Trade Organization (IDE-JETRO). The project was titled "Crafting Japan-
U.S. Water Partnerships: Promoting Sustainable River Basin Governance in China" and
was generously funded by the New York office of the Japan Foundation's Center for
Global Partnership (CGP). Within this project the tri-national research team members
wrote research papers that were published in IDE Spot Survey No. 28 in March 2005
*Promoting Sustainable River Basin Governance: Crafting Japan-U.S. Water Partnerships
in China*. Also in November 2005, a special issue of *Ajiken World Trend*—an IDE-
JETRO analytical journal (in Japanese) that explores the future prospects of developing
countries—featured our Japanese team members, whose papers focused on international
cooperation for sustainable river basin governance in China. This report incorporates
some of the research published in the IDE Spot Survey, but includes predominantly
new information and updates. The Chinese and Japanese versions in this print version
summarize the main points of the English text. Full translations of the English text
in Chinese and Japanese are available on the China Environment Forum Web page at
www.wilsoncenter.org/cef under the publication link.

Many individuals and organizations deserve our thanks for helping us create this
publication. First and foremost were the research team members who enthusiastically
participated in three intensive study tours on river basin governance in China, Japan, and
the United States. The ten members were: Carol Collier, Kaori FUJITA, HU Kanping,
Naoki KATAOKA, Reiko NAKAMURA, Richard Volk, WANG Yahua, Gary Wolff,
Nanae YAMADA, YU Xiaogang. They used these study tours to design and carry out
individual research papers aimed at exploring how China could pursue integrated river
basin management by studying lessons from Japan and the United States in three crucial
areas: river basin management institutions, financing, and public participation. Another
vital member of the team was the CEF project assistant Timothy Hildebrandt who was
key in conceptualizing and organizing the study tour work crucial for the success of this
project. We wish also to thank Mikiyasu NAKAYAMA, who contributed a valuable research paper on China's international transboundary rivers to our first publication. Our
research team also greatly benefited from insights provided by Masahisa NAKAMURA
(Lake Biwa Research Institute) and Naohiro KITANO (Kyoto University) who presented and commented on the team's research at an international workshop we held 7
October 2005 in Tokyo.

During the study tours countless people sat down with our group to share their experience and insights on how to manage river basins more effectively. We therefore owe
thanks (in alphabetical order) to individuals in the following organizations for assisting

our research team: Asaza Fund, Chesapeake Bay Foundation, Chesapeake Bay Program, China Council for International Cooperation on Environment and Development, China Environment and Sustainable Development Reference and Research Center, Chinese Ministry of Water Resources, Conservation International, Delaware River Basin Commission, Embassy of Sweden in Beijing, Embassy of Switzerland in Beijing, European Union office in Beijing, *Green China Times, Green China Journal*, Green River, Green Watershed, GTZ, Hai River Conservancy Commission, Infrastructure Development Institute (Japan), Institute of Developing Economies, Interstate Commission on the Potomac River, JBIC, JICA, Japan Water Agency, Kanagawa Prefecture Tax Reform Office, Maryland Department of Natural Resources, Momoyama Gakuin (St. Andrew's) University, New York Regional Plan Association, People's University, Pacific Institute, Ramsar Center Japan, Tianjin Environmental Protection Bureau, Tokyo Keizai University, Tokyo Metropolitan Local Office of Bureau of Ports and Harbors, Tsinghua University, Tsukuba University, UK's DFID, University of Tokyo, U.S. Army Corp of Engineers, U.S. Environmental Protection Agency, Wetlands International, Woodrow Wilson Center, World Bank Beijing office, World Fish Center, and WWF-China.

We also are indebted to a number of people who supplied us information or graciously read early drafts of this brief and provided us with invaluable suggestions to strengthen and clarify our ideas. Baruch Boxer, Patrick Freymond, Ping Hojding, Bryan Lohmar, Kaori Fujita, Jim Nickum, Richard Volk, Wang Yahua, Naoki Kataoka, Wen Bo, Nanae Yamada, Ma Jun, Fengshi Wu, Mikio Ishiwatari, and Naoki Mori. We must thank Linden Ellis, Charlotte MacAusland, Louise Yeung, and Lulu Zhang for helping us compile facts and enduring the time-consuming editing work. We also greatly appreciate Serena Lin's excellent translation work. Lastly, we are grateful to the unflagging support from staff at the CGP who challenged us to make this a productive project, attended portions of the study tours, and even provided meeting space for our large conference in Tokyo. At CGP we would particularly like to thank Carolyn Fleisher, Hara Hideki, Jun'ichi Chano, and Atsuko Sato. While these individuals and others at the Wilson Center and IDE made critical contributions, we remain responsible for all the report's content. The views expressed in this report are those of the authors alone and not necessarily of the Wilson Center and IDE.

EXECUTIVE SUMMARY

China is facing numerous water crises—lakes and rivers contaminated with toxic pollutants from unregulated industries and untreated urban wastewater; severe water shortages stemming from over pumping of ground and surface water; and flood disasters caused by deforestation and destruction of wetlands. Water degradation and scarcity in China contribute to population movements, health risks, and food security problems. Water problems ultimately have the potential to affect China's social, economic, and political stability.

At the core of China's water challenges is the need to protect the country's river ecosystems. The need to mitigate the threats to China's rivers has catalyzed domestic and international efforts to strengthen laws, policies, and projects to promote integrated river basin management (IRBM) and more holistic pollution prevention strategies. One central strategy for implementing IRBM has been the Chinese government's attempts to reform the river basin commission system. These top-down measures are crucial for true reform in river

Ganjiang River scene
Photo Credit: Xiao Qiping

management, but equally important will be greater empowerment of citizens and non-governmental organizations (NGOs) to participate in the decision-making and monitoring of river development and protection. A few international environmental NGOs have set up river basin protection projects in China that have brought together government agencies, communities, and Chinese NGOs to create multi-stakeholder projects to protect local rivers.

The governments and NGOs in the United States and Japan are independently undertaking some water and river protection projects in China. However, many of these projects are small-scale and short lived, which limit their ability to promote needed institutional change for true IRBM in China. In order to have a greater impact on promoting IRBM in China, the U. S. and Japan could jointly pursue initiatives in the areas of watershed management, financing, and stakeholder participation.

This report aims to present some options for the government, NGO, and research sectors in the U.S. and Japan (as well as other countries) to undertake collaborative river basin governance projects in China. To set the stage for a discussion of greater international cooperation around water in China, Part I discusses the magnitude of water problems in China. Next, Part II reviews the effectiveness of current water laws and

Sulfur and other runoff from factories—like these on the bank of Dadu River, a branch of the Yangtze—exacerbate the bigger water quality problems of erosion, sewage, and garbage in the Yangtze. Photo Credit: Yang Xin

institutions, as well as the small, yet growing indigenous NGO activity on water protection issues in China. Part III presents an overview of international aid and assistance in China to promote sustainable water management, as well as highlight the gaps in this work. The conclusion in Part IV provides some potential themes that Japanese and U.S. governments, NGOs, and research communities could pursue jointly (or in parallel projects) in China to promote sustainable river basin governance.

PART ONE:
CHINA'S RIVERS IN CRISIS

Over the past 25 years, the Chinese economic miracle has brought millions out of poverty, but at a cost to China's environment. The statistics on China's environmental problems highlight a potentially grim outlook for the country. Sixteen of the world's twenty most polluted cities are in China; China already consumes more energy (most of it low-grade coal) and emits more greenhouse gases than any country except the United States, and may surpass the United States in both categories within two decades; acid rain from coal burning affects two-thirds of the country (as well as Korea and Japan); twenty percent of the country's plant and animal species are endangered; water scarcity in the northern region has created eco-refugees fleeing farmland turned desert; and more than 75% of the rivers flowing through Chinese cities are unsuitable for drinking or fishing.[33] Even the Vice Minister of China's State Environmental Protection Administration (SEPA) Pan Yue has stated that the magnitude of China's environmental degradation— which costs the country approximately 8% of its annual GDP growth—has made the economic miracle more of a myth.[2] Among the many environmental problems, severe water scarcity, growing water pollution, and mismanagement of river ecosystems represent major threats to economic, ecological, and human health in China.

China's Water Woes

The Chinese government has been increasingly prioritizing water conservation and pollution control, however, the speed of economic growth, population pressures, and lack of law enforcement at the local level have meant progress addressing these pressing water problems has been slow. (See Box 1 for some water statistics).

Water Scarcity
China's annual per capita water supply is 25% below global average and by 2030 per capita supply is expected to fall from 2,200 cubic meters (m^3) to below 1,700 m^3, a level that meets the World Bank's definition of a water-scarce country.[3] Water scarcity is most acute in north China where per capita water resources are only 750 m^3 per year.[4] While agriculture still consumes nearly 80% of water resources in China, water consumption in industrial and domestic sectors has been rising quickly. These three thirsty sectors and lack of conservation measures are accelerating the depletion of water resources, particularly in the dry north where despite only having 24% of China's water resources, produces grains that account for more than 45% of China's GDP.[5]

Excessive water withdrawals and land degradation in northern and western China have caused desertification to advance at an annual rate of 1,300 square miles, affecting 400 million people.[6] Twenty-four thousand villages in northern and western China have been abandoned or partially depopulated due to growing desertification that has made farming untenable.[7] This desertification also has exacerbated the spring sandstorms—100 are expected between 2000 and 2009, a marked increase over the 23 in the previous decade.[8] These sandstorms not only affect China, Korea, and Japan, but also reach the U.S. west coast.

The annual costs of water scarcity and pollution on agricultural losses range from the World Bank's high $24 billion estimate to Chinese news media quotes of $8.2 billion.[9] While most severe in the north, water scarcity has become a major obstacle to sustainable development throughout the country. Four hundred of China's 640 major cities face water shortages, which in 2003 cost $28 billion in lost industrial output.[10] In China's rural areas approximately 60 million people face challenges in getting enough water for their daily needs.[11] The growing magnitude of water scarcity is illustrated quite starkly in the Yellow River, which since the mid-1990s has grown so dry the river often does not reach the ocean for up to 200 days a year.[12]

Notably, increasing water supply through major dam and water diversion projects continues to be a cornerstone of China's response to water shortage. These huge projects—particularly dams—are costly and increasingly the target of opposition by local residents who stand to lose their land and livelihoods. Some studies estimate that more rigorous water conservation efforts could save China 100 to 200 billion cubic meters of water per year and thereby cut China's current water consumption about one quarter, obviating the need for such a large number of expensive and increasingly controversial massive dam and diversion projects.[13]

Water Pollution

Severe pollution is affecting all major rivers in China threatening human health and disrupting industrial production, as well as destroying river ecosystems. Weakly regulated industries and insufficiently coordinated management of water resources are two of the main institutional failures that are driving this severe water pollution problem in China. Since 2002, approximately 63 billion tons of wastewater flow into China's rivers each year, of which 62% are pollutants from industrial sources and 38% are poorly treated or raw sewage from municipalities.[14]

Wastewater treatment was a major priority in the Tenth Five-Year Plan (2001-2005), however, a 2004 inspection by SEPA of sewage treatment plants built since 2001 found that only half of them were actually working and the other half were closed down because local authorities considered them too expensive to operate.[15] At the end of 2002, the official municipal wastewater treatment rate was 39.9% and in rural areas these rates are much lower.[15]

Perhaps most telling of the weaknesses in protecting China's waters is the case of the Huai River, which despite a decade-long central government campaign that began in 1993, is still one of the most polluted in China. The highly polluted Huai is linked directly to high cancer rates and other serious human health impacts in the basin. For example, for many years no young men from certain villages in the Huai River Basin have been healthy enough to pass the physical examination required to join the army.[16]

BOX 1:
TRENDS OF WATER USE, SHORTAGES AND POLLUTION IN CHINA

Water Use

- Annual demand for water is expected to triple from 120 to 400 billion tons during 1995-2030.
- From 1980 to 2003 agricultural consumption (including forestry and wetlands) declined from 83.4% to 64.5%, while industrial and household consumption increased from 10.3% to 22.1%, and from 2% to 11.9%, respectively.[19]
- Between 1980 and 2000, Chinese urbanites increased per capita daily water consumption about 150 percent–from less than 100 liters in 1980 to 244 liters in 2000.[20]
- Only 43% of the water consumed in agriculture is used efficiently for irrigation, compared to 70% to 80% in developed countries.[21]
- China loses as much as 25% of the water transmitted through pipes due to leaks which is considerably higher than the 9% lost in Japan or 10% in the United States.[22]
- Groundwater comprises 30 percent of the China's total urban water supply, but due to environmental problems caused by excessive extraction (e.g., less water to dilute pollutants), only 63% of the urban areas have groundwater that is potable without treatment.[23]
- In 2002 the amount of water used for every $10,000 worth of GDP in China was 537 m[23], four times the world average and 10 to 20 times that in developed countries.[24]

Water Scarcity

- China's annual water shortage ranges from 30 to 40 billion m[24], of which the urban water shortage is 6 billion m[24].
- In 2003, floods and droughts led to economic losses of $24 billion; while desertification cost the country $6 billion.[25]
- SEPA estimates that in 2003 and 2004, water scarcity cost China $28 billion in lost industrial output.[26]
- Over the coming decades desertification could cause thirty to forty million Chinese farmers to migrate due to lack of access to arable land and water.[27]
- Water shortages in the North China Plain threaten China's goal of food self-sufficiency, for the region produces more than 50% of the nation's wheat and 33% of its maize.

Water Pollution

- Nearly 700 million Chinese lack access to safe water and consume water contaminated with animal and human waste that exceeds the maximum permissible levels for fecal coliform bacteria.[28]

(continued)

With a low wastewater treatment rates (40% nationwide) cities are a major source of pollutants flowing into China's rivers. Photo Credit: Xiao Qiping

- Underground water in 90% of Chinese cities is polluted, which raises human health concerns since 70% of the Chinese population depends on underground water for drinking.[29]
- In 2003, industrial and domestic wastewater emissions totaled 69 billion m³, double the 1980 level. Each year one-third of industrial wastewater and two-thirds household sewage are emitted untreated.[30]
- Over 50 percent of the length of the Hai River, one of the major rivers in northern China, has worse than level V water quality (poor).[31]
- Water pollution cost China's fisheries $130 million in 2004, an increase of over $40 million from the previous year.[32]
- Along China's major rivers—particularly the Huai, Hai, and Yellow—communities report higher than normal rates of cancer, tumors, spontaneous abortion and diminished IQs due to the high level of contaminants in the soil and water.[33]

Degraded Water Ecosystems

While pollution and over extraction are the major causes of degraded watersheds in China, many rivers—particularly the Yangtze—are equally threatened by deforestation, conversion of wetlands for agriculture, and unsuitable infrastructure projects in the flood plain, all of which have led to bigger and more damaging floods. Moreover, ill-planned hydrological projects on the Yangtze have disrupted the river's natural flow, damaging the basin's ecosystems and leading to considerable loss in biodiversity and the productivity of the river.

In addition to the environmental and economic costs of water degradation, polluted water and shortages have contributed to social unrest in China. The Western news media and nongovernmental community tend to focus on high-profile water conflict stories, such as the problems of citizens relocated for the construction of the Three Gorges Dam. However, inter-provincial disputes and lower-level conflicts are growing in number and even becoming violent. (See Box 2).

Three Core Elements of IRBM

Examination of the problems facing China's rivers opens up an opportunity to understand the country's growing water crises, as well as the political and social problems hindering effective river basin governance. While nearly all countries in the world face multiple challenges in protecting water resources, China is seriously failing to sustainably manage its water resources, particularly rivers. The detrimental impacts of uncontrolled development and insufficiently coordinated water management institutions underscore the need for China to adopt a more holistic approach to river management—specifically integrated river basin management (IRBM). Within the complex IRBM concept we believe there are three key institutions that Chinese policymakers, NGOs, and international donors should first emphasize to promote better river basin governance in China: (1) river basin management institutions, (2) financing mechanisms, and (3) public participation. Below is a short review of these three institutions in China.

Fragmented Management Institutions

Ineffective and insufficiently coordinated management to protect water resources lies at the core of China's water problems. China's first Water Law (passed in 1988) and supporting regulations mandated water conservation efforts (such as water fee collection, allocation programs, water use permits, installation of water efficient equipment). However, weak monitoring and enforcement capability at the local levels and within river basin commissions, as well as difficulties in creating clear water-use rights hinder many of these water management reforms. A core goal of the 2003 amended Water Law was to empower river basin commissions in order to improve the implementation of water conservation and management measures.

China's seven river basin commissions (RBCs) were initially created in the 1950s to exploit water resources, generate electricity, mitigate flood damage, and provide facilities for navigation. As branches the Ministry of Water Resources, RBCs possess strong technical and hydrological expertise but often lack the management capacity to

BOX 2:
WATER CONFLICTS IN CHINA

The contentious nature of managing and protecting water is aptly captured in a quote attributed to Mark Twain: Whisky is for drinkin' and water is for fightin'. While *baijiu* and not whisky is the firewater of choice in China, Chinese government agencies, provinces, counties, villages, and individuals do fight over access to clean water or over damages from polluted water. In recent years the Chinese government has become more transparent in revealing numbers on protests related to environmental issues.

- The director of the Policy and Regulatory Department of China's Ministry of Water Resources (MWR) Gao Erkun, reported in a July 2003 meeting that from 1990 to 2002 over 120,000 water quantity conflicts had been reported to the ministry.
- In the summer of 2005 the central government, for the first time, announced that in 2004 3.76 million Chinese, mostly disadvantaged groups, took part in 74,000 mass protests. Many of these protests are sparked by land grabs by local governments, closing factories, and increasingly, environmental pollution.[34]
- In the mid-1990s, a Central Committee of the Chinese Communist Party report acknowledged that environmental degradation and pollution represented one of the four leading causes of social unrest in the China.[35]

Citizen Protests

Environmental problems, particularly lack of access to clean water has sparked a growing number of citizen complaints and protests since 1997. In 1997, citizens sent 16,758 letters about pollution to environmental bureaus, which increased fivefold to 60,815 in 2003—noise pollution was the top complaint, with air and water pollution being the next most common complaints.[36] One long-standing water dispute began in the 1980s when villages along the Zhang River (a tributary of the Hai in northern China) undertook near guerilla warfare destroying each other's water diversion canals after a growing number of government-sponsored water diversions further upstream created a severe water shortage in the basin. Local governments in the basin have been unable to resolve these conflicts and only in the past few years has a major MWR initiative to mediate the conflicts begun to calm the situation.[37] A number of water pollution protests have emerged in the wealthy province of Zhejiang. In April 2005, 60,000 people in the village of Huaxi, Zhejiang, protested against an industrial park in which 13 chemical plants had been polluting the water and soil around the village.[38] After police began beating elderly citizens blocking the entrance to the park, farmers from surrounding villages arrived to drive the police away, two people were reported killed in the violence. The local government promised to close and move the factories, but as of August 2005 nothing had been done and villagers threatened new protests. In July, 2005, more than 10,000 people in Zhejiang province protested the toxic emissions from a pharmaceutical plant that was contaminating land and water and harming public health.[39]

Transboundary Conflicts

In terms of transboundary water issues, China's development of the upper reaches of the Mekong River causes much concern in the region. As only an observer rather than a full member of the Mekong River Commission, China is not obligated to clear the planned eight dam construction projects with downstream countries. Pollution from transboundary rivers represents another major source of tension with surrounding countries, exemplified in the major benzene spill into the Songhua River in November 2006 that impacted both Chinese and Russian cities.

Conflict Resolution in the Courts

Not all conflicts over water projects turn violent. The government, NGOs, and private law firms are making efforts to create channels for peaceful dialogue on water problems. The Chinese government acknowledges the necessity of improving water protection to prevent conflicts, as well as strengthening water conflict resolution mechanisms. Within the National Water Law there are provisions for administrative arbitration when conflicts arise between government jurisdictions. Although MWR employs 60,000 people to deal with water quantity conflicts, administrative arbitration methods do not always work because local protectionism prevents local water and environmental bureaus from enforcing judgments.[40] In the water pollution sphere, SEPA has set up a third-party mediation system within the seven major river basins to oversee water management problems and conflicts. SEPA also is currently drafting a policy to address cross-provincial water pollution disputes.

Beginning in the late 1990s private Chinese law firms began specializing in environmental cases, many of which have been class-action water pollution cases. These private lawyers have been winning cases for pollution victims by moving cases to courtrooms outside the influence of the local government where the water conflict took place.[41] In January 2005 a landmark settlement over water pollution took place in Inner Mongolia, when two Chinese paper mills and a local water-treatment bureau agreed to pay $285,000 to the Baotou City Water Supply Company. In 2004 polluted wastewater from the two paper mills forced the water supply company to stop using water from the Yellow River for five days, which led to over $300,000 in economic losses for the company.[42] This case highlights how Chinese courts could potentially become an effective tool for pollution victims to use to gain justice and compensation.

Since the mid-1990s, Chinese journalists have been quick to report on water pollution and other incidents of environmental degradation, giving the news media a growing role in channeling information on environmental threats to the public and checking polluters. (Photo Credit: China Environment News)

monitor and enforce water protection and conservation measures. Generally, Chinese RBCs do not focus enough on the ecological health of rivers or sufficiently address ways of balancing upstream and downstream water needs. The effectiveness of river commissions is also limited due to poor (and sometimes even adversarial) relationships with other government agencies, as well as provincial and local governments. Another major shortcoming of Chinese RBCs is that they fail to incorporate broad stakeholder input in basin management efforts.

Limited Financing Tools

In river management planning and projects to divert water resources, river basin commissions and other government agencies do not sufficiently consider economic, social, and environmental costs. Although the central government sets lofty river and water clean up goals in campaigns and in five-year plans, the funding often falls short of what local governments need. For example, the ten-year Huai River campaign had little impact on the river. Moreover, although there were numerous water clean up targets in the Tenth Five-Year Plan, the government fell 30% short on its funding for environmental goals in the plan.[17]

Instead of depending on central government subsidies, China needs to provide economic incentives for industries and local governments to protect river ecosystems, particularly downstream users compensating upstream protection measures. Chinese RBCs and cities lack financing mechanisms such as revolving funds and bonds, which could fund the construction of sorely needed wastewater treatment facilities. A major obstacle to such mechanisms is the low rate and low collection levels of water fees. Market tools such as green taxes, water trading, and upstream-downstream compensation strategies, which could promote conservation of river resources have been slow to develop in China.

Lack of Transparency and Public Consultation in Water Policy Decision-making

At both central and local levels, governments do not inform and consult with citizens about proposed development projects and water resource management initiatives. Thus public participation in the water policy sphere is generally limited to complaints and protests. Chinese citizens are allowed to make formal complaints to the government on damages from pollution and to sue polluters (and recently even government agencies), however these efforts do not always effect change. Greater citizen and NGO involvement in monitoring water policies and projects holds promise of improvements in protecting China's river basins.

Potential of a U.S.-Japan Water Partnership in China

The economic reforms in China that have brought rapid industrialization, raised standards of living, and freed many rural people from agricultural work have produced declining environmental conditions that directly impact the health of the Chinese people and their economy. The need to build U.S., Japanese, and Chinese environmental partnerships on multiple levels is particularly crucial as China's integration into the world economy speeds up both economic growth and environmental degradation. Although China's water challenges are severe, they do hold many opportunities for U.S.-Japan assistance in the areas of watershed management, financing, and stakeholder participation.

Notes

1. Economy, Elizabeth. (2004). *The River Runs Black: The Environmental Challenge to China's Future*. New York: Cornell University Press. Baldinger, Pamela and Jennifer Turner. (2002). *Crouching Suspicions Hidden Potential: U.S. Environmental and Energy Cooperation with China*. Washington, DC: Woodrow Wilson Center.

2. "The Chinese Miracle Will End Soon." (2005, March 7). *Der Speigel*.

3. Economy, Elizabeth. (2004, September 22). *China's Environmental Challenges*. Testimony for Congressional Executive Commission on China, Washington, DC.

4. U.S. Embassy in Beijing. (2003). *China's Water Supply Problems*. [On-line]. Available: http://www.usembassy-china.org.cn/sandt/ptr/water-supply-prt.htm

5. Lohmar, Bryan, Jinxia Wang, Scott Rozelle, Jikun Huang, and David Dawe. (2003). *China's Agricultural Water Policy Reforms: Increasing Investment, Resolving Conflicts, and Revising Incentives*. (Bulletin Number 782). United States Department of Agriculture, Economic Research Service Agriculture Information.

6. *China Daily*. (2002, July 1). "China Launches Unprecedented Forestry Programs." [On-line]. Available: www.china.org.cn/baodao/english/newsandreport/2002july1/13-2.htm

7. Ibid.

8. Lubick Naomi. (2005). "China's Changing Landscape." *Geotimes*, October, 18-21.

9. World Bank. (1997). *Clear Water Blue Skies*. Washington, DC: World Bank.

10. "Pollution Costs China's Fisheries US$130 million in 2004." (2005, May 23). *Asia Pulse*.

11. UNDP. (1999). *The China Human Development Report*. (New York: Oxford University Press).

12. Wang Yahua. (2005). "River Governance Structure in China: A Study of Water Quantity/Quality Management Regimes. In Jennifer L. Turner and Kenji Otsuka (Eds.), *Promoting Sustainable River Basin Governance: Crafting Japan-U.S. Water Partnerships in China*. IDE Spot Survey No. 28 (pp. 23-36). Chiba, Japan: Institute of Developing Economies/IDE-Jetro. [Online]. Available: http://www.ide.go.jp/English/Publish/Spot/28.html

13. U.S. Embassy in Beijing. (1997). "PRC Water: Waste a lot, have not." [Online]. Available: www.usembassy-china.org.cn/english/sandt/index.html. Nickum, James. (1998). "Is China Living on the Water Margin?" In Richard Louis Edmonds (Ed.), *Managing the Chinese Environment* (pp. 156-173). Oxford: Oxford University Press.

14. U.S. Department of Commerce, International Trade Administration. (2005). *Water Supply and Wastewater Treatment Market in China*. Washington, DC.

15. "China Politics: Green-tinted Glasses." (2004, July 6). *The Economist Intelligence Unit*.

16. U.S. Department of Commerce. (2005).

17. Economy, Elizabeth. (2004). *The River Runs Black*.

18. Otsuka, Kenji. (2005). "Water Pollution Control in Huai River to be Re-evaluated. " *Ajiken World Trend*, Vol.112, January, 36-39. [In Japanese].

19. *Economic Daily*. (2005, August 8). "Prioritizing Water Conservation: An Interview with Luliang Qi."

20. Dabo Guan and Klaus Hubacek. (2004, October 7-8). *Lifestyle Changes and its Influences on Energy and Water Consumption in China*. Paper presented at 6th Conference for postgraduate students, young scientists and researchers on Environmental Economics, Policy and International Environmental Relations, Prague.

21. Ibid.

22. "China's Water Slows to a Trickle." (1996, July 30). *Financial Times*, page 5.

23. "Chinese Cities Facing Water Crisis." (2003, November 3). *China Daily*. [Online]. Available: http://www.china.org.cn/english/environment/79010.htm

24. *Economic Daily*. (2005, August 8).

25. "Country Sustains Weather of Extremes." (2004, June 29). *China Daily*. "Farmers Need More Help to Fight Desertification." (2004, June 23). *Business Daily Update*.

26. Economy, Elizabeth. (2005, October 28). "China's Environmental Challenge." Paper presented at Eurasia Group Workshop, New York. "The Situation and Problems of the Use of Water Resource in China." (2004, July 13). *Economic Daily*. [In Chinese].

27. Vaclav Smil. (1994).

28. Stockholm Environment Institute and UNDP. (2002). *China Human Development Report 2002*. New York: Oxford University Press.

29. Chao Shengyu. (2005, December 28). "Underground Water in 90 Percent of Chinese Cities Polluted." *Jinhua News Network*.
[Online]. Available: http://www.jhnews.com.cn/gb/content/2005-12/28/content_552266.htm

30. "Key to Ease Water Shortage in China is to Establish a Society of Water Conservation." (2004, December 21). *Economic Daily*. [In Chinese].

31. U.S. Embassy in Beijing. (2003) and SEPA. (2004). "Public Information of China Environment." [Online]. Available: http://www.zhb.gov.cn

32. "Pollution Costs China's Fisheries US$130 million in 2004." (May 23, 2005). *Asia Pulse*.

33. Economy. (2004). *River Runs Black*.

34. Elkington, John and Mark Lee. (2005, August 23). "China Syndromes: Will Hard-won Environmental and Social Gains Survive China's Econo.mic Rise?" *Grist Magazine*. [Online]. Available: http://www.grist.org/biz/fd/2005/08/23/china/index.html

35. Shi Ting. (2004, September 9). "Academics Warn that Social Unrest Could Post Threat to Economy; Corruption is Singled out as a 'Highly Possible' Trigger." South China Morning Post.

36. SEPA. (2004). *China Environmental Yearbook*. Beijing: SEPA. [In Chinese].

37. Hildebrandt, Timothy and Jennifer Turner. (2005). "Water Conflict Resolution in China. *China Environment Series*, Issue 7, 99-103.

38. "Sixty Thousand People Protest Against Pollution." (2005, April 14). *AsiaNews.it*. [Online]. Available: http://www.asianews.it/view_p.php?l+en&art+3036

39. Economy, Elizabeth. (2005, October 28).

40. Hildebrandt, Timothy and Jennifer Turner. (2005).

41. Ma Jun. (2003). "Sue you Sue me Blues." *China Environment Series*, Issue 6, 89-98.

42. Zhou Yang. (2006, January 5). "Chinese Firms Pay in Pollution Deal." *The Washington Post*, p. D6.

PART TWO:
DOMESTIC INITIATIVES ON
RIVER BASIN PROTECTION

The difficulties in effectively addressing the growing water scarcity and pollution problems have led the Chinese government to adopt new (and update old) laws addressing water management and pollution problems, as well as reform the river basin commission system. Water protection also has become a priority investment area in the two most recent five-year plans, for sustainable water supplies are crucial to fuel economic growth. These top-down measures are critical for reforming the water management laws and institutions and improving the water pollution prevention infrastructure. Equally important will be for the government to continue expanding the political space for bottom-up citizen and nongovernmental organization (NGO) involvement in the water sector. Below we first briefly introduce the scope and impact of the Chinese government's investments, laws and institutions to manage and protect the country's endangered water resources. We then highlight some openings for public participation in protecting water resources, particularly NGO activities.

Top-Down Initiatives Addressing China's Water Woes

High-Level Prioritization

Under the Tenth Five-Year Plan (FYP, 2001-2005), the government set admirable environmental protection targets—particularly around water—but fell short by 30% of the promised $85 million investment. Under the Tenth FYP, the government aimed to construct or expand 145 urban wastewater treatment plants in the basins of the Huai, Hai, and Liao rivers and the Tai, Chao and Dianchi lakes. In addition, urban sewage treatment rates were supposed to reach 50 percent.[1] Although the Chinese government made progress in increasing the rates of industrial and municipal wastewater treatment, a SEPA inspection of sewage treatment plants built during the Tenth FYP revealed only half of were actually in operation.[2]

During this FYP period the government did continue enforcement of a logging ban in southwest China, as well as investment into a program to encourage the conversion of cropland to forests on sloped fields. Both these programs have helped alleviate the serious erosion that exacerbated the 1998 flooding of the Yangtze River.[3] In late 2005, SEPA announced that water pollution control efforts in China have led to some improvement of water quality.[4] However, such claims may overstate the progress. For example, during the ten-year campaign to clean up the Huai River, SEPA reported great improvements in water quality, but at the conclusion of the campaign was forced to admit the river was (and remains today) severely polluted. One high-level Chinese water expert stated that it was impossible for the Tenth FYP goals of cleaning up the Huai River to be met by the end of 2005.[5]

The Chinese Communist Party issued their proposal for the Eleventh FYP (2006-2010) in early October 2005 and the final plan will be approved the by National People's Congress in March 2006. The current draft contains statements expressing very holistic principles of protecting the natural ecosystems and conserving energy. One part of the plan notably calls for environmental policies to be made to protect wetlands and repair damaged ecosystems along China's coastlines. The FYP also includes exhortations to intensify pollution control in three key rivers (Huai, Hai, and Liao) and three major lakes (Tai, Chao, and Dianchi), as well as the Three Gorges Dam area and the upper reaches of the Yangtze and Yellow rivers. The plan also prioritizes pollution control along the route of the South-North Water Transfer project.[6] Under the plan SEPA aims to raise the urban wastewater treatment rate from the current 45 percent to 60 percent in all cities.[7] Former SEPA Minister Xie Zhenhua noted that total investment in environmental protection during the Eleventh FYP is aimed to reach over 1,300 billion Yuan ($156.6 billion), more than 1.5 percent of the country's GDP.[8]

Water Quantity Management in China

Although China has formulated a fairly comprehensive array of water protection laws, major institutional problems are hindering effective management of water resources, particularly rivers. One core challenge to water quantity management is the lack of a clear water rights system, for water belongs to the state and has therefore been treated as an open access resource. Unclear water rights have discouraged conservation, as well as water trading, which could allow water to be transferred without conflict between sectors (particularly from agriculture to industry).

The competition among various bureaucracies to control water resources in China—often referred to as the many-headed dragon (*Duo Longtou*)—is the other major institutional driver of poorly integrated management of water. The main combatants in controlling water are the Ministry of Water Resources (MWR, responsible for water quantity issues) and the State Environmental Protection Administration (SEPA, responsible for overseeing water pollution control). While the ministries of agriculture and construction and the Bureau of Fisheries also lay claim to various aspects of water management, overall MWR possesses the most power to make decisions over water quantity and river basin management throughout China. Notably, as China's political system has been decentralizing over the past 25 years, in the area of water the central government has been re-centralizing control of rivers under MWR and its seven major river basin commissions.

China passed its first National Water Law in 1988, which attempted to clarify authority over the management of water and create the framework for some water conservation institutions to better define water rights. For example, the law included: (1) a requirement for increasing water fees; (2) a water withdrawal permit system to assign water user rights to industries, farms, and cities, which in turn will enable local governments to collect more water fees; and (3) a water runoff allocation scheme to divide water rights among provinces in watersheds under stress.[9]

Local governments have been major hindrances to the long-overdue fee and permit reforms due to fears that limiting water use will hurt the local economy. However, in recent years a number of major cities have begun to increase water fees and install more water meters, which are key changes needed to slow the dangerous overdrawing of surface and groundwater resources. Nevertheless, when cities lack water they usually opt to tap new supplies rather than strictly enforce fee, permit, or other conservation policies. Beijing municipality pursues this supply side management to quench the capital's growing thirst.[10]

One very positive trend to improve water management at the local level has been the creation of municipal Water Resource Bureaus in a handful of cities. This new bureau brings together local environmental, water, and construction bureaus to jointly manage water resources—a local innovation trying to deal with the many-headed dragon problem. Another promising, albeit not yet legal, development coming from the local areas has been experiments in trading water between counties or between industrial and agricultural users. The first water trade took place in Zhejiang province in 2000 when Yiwu county bought permanent water use rights for 50 million m^3 per year of reservoir water from the upriver Dongyang county.[11]

By the mid-1990s, after considerable pressure and effort by MWR and its local water bureaus, the permits for water withdrawals were issued in many regions, but the amounts permitted often did not push much needed water conservation and were not always well enforced. The water allocation scheme that has been introduced in few basins also was not completely successful because: (1) MWR and provincial governments did not create enforcement institutions, and (2) river basin organizations lacked sufficient clout and ability to coordinate and monitor such water allocations.[12]

The 2002 amendment of the Water Law aimed to remedy some of the shortcomings of the previous law and address the growing water shortages by requiring river basin commissions (RBCs) to allocate water among all the provinces within the basin following a total amount control strategy (e.g., viewing river water holistically and reserving some water for ecological flows). A strictly enforced allocation system is meant to push local governments to monitor and enforce water withdrawal permits and prioritize water conservation. Managing basin-wide allocation schemes is difficult logistically and politically, which is why among the seven major RBCs only the largest and most powerful one—the Yellow River Conservancy Commission (YRCC)—has formulated and implemented a fairly successful runoff allocation scheme.

Challenges to Water Quality Management

In 1984 the Water Pollution Prevention Law (WPPL) was passed and revised in 1996. This law states that environmental protection bureaus at each level of government are empowered to manage and monitor inspections to prevent water pollution. Despite possessing the legal authority to regulate pollution, local protectionism of industry has prevented environmental protection bureaus (EPBs) from effectively using WPPL to prevent water pollution. For example, although pollution charges are low the local governments will often return up to 80 percent of the charges back to the polluting industries.[13] Local officials frequently turn a blind eye to illegal discharges of pollutants from an industrial firm that pays a large amount of tax to the local government coffers.

Some inconsistent stipulations in the WPPL and Water Law have aggravated interministerial conflicts and thereby complicated efficient implementation of water quality planning, protection, and monitoring.[14] Although SEPA regulates water quality through

BOX 3:
CHINESE ENVIRONMENTAL NGOS ENGAGED IN WATER RESOURCE PROTECTION

Ganjiang Environmental Association (Jiangxi Province)

Over the past decade many highly polluting industries have relocated to Jiangxi province, which has caused the water quality of the Gan River (Ganjiang) to dramatically deteriorate. Cancer rates along the Ganjiang appear to be increasing as well. In response to this rapid degradation of the river, in 2003 concerned citizens and environmental experts created the Ganjiang Environmental Association. Led by Xiao Qiping with a four-member staff, the association has been: (1) conducting water quality research, (2) producing publications on water resource protection, (3) sponsoring photo exhibitions and lectures at schools, and (4) shooting a documentary on environmental protection needs in the basin. Beginning in July 2003 members of the association began taking motorbike trips to survey the pollution problems in the middle reaches of Ganjiang, compiling 15-hours of video footage and thousands of pictures. Local and national news media have used some of these photos to publicize the river's pollution problems and the association's work, which has helped this NGO gain greater influence. Since 2003, the association's monitoring of the river has uncovered pollution violations by enterprises in four townships in the basin (Taihe, Futan, Xiajiang, and Fengning).

Green Hanjiang (Hubei Province)

Green Hanjiang registered in September 2002 in Hubei province and is the first environmental group working on the Han River (Hanjiang). This NGO aims to promote environmentally friendly development around the Hanjiang through public education and information dissemination. Green Hanjiang's main activities include doing research on environmental hotspots around Hanjiang, communicating public concerns to local government agencies, acting as a watchdog against local pollution, and educating rural residents on the importance of sustainable development. This NGO also has advocated for greater compensation for citizens who will be displaced by the construction of the South-North Water Transfer project.

Green River (Sichuan Province)

Green River was created by nature photographer Yang Xin, who wished to better inform policymakers on how to protect the fragile headwaters of the Yangtze River by promoting better science and study of the region. The Sichuan-based Green River has worked since 1994 to protect the ecologically fragile Yangtze headwaters region through activities at two ecological research centers. Core Green River projects include: (1) cooperation with local scientific research organizations and journalists to survey and research the quality of the Yangtze River headwaters in order to accumulate baseline data on the health of the river and to help design an effective environmental protection plan for the upper basin; and (2) recruitment of volunteers to educate local rural communities and tourists about the threats to the Yangtze River ecosystem. In a new initiative, Green River is developing a program to help promote ecologically sustainable tourism in one Tibetan village in the Minjiang Basin (a tributary of the Yangtze). Besides educating villagers on ecotourism and environmental protection, in 2006 Green River will build some waterless toilets and an ecologically friendly solid waste treatment facility in one village.

Green Watershed (Yunnan Province)

Green Watershed is an NGO focusing on integrated watershed management in the Lancang-Mekong River Basin in Yunnan province. Founded in 2002, the mission of Green Watershed is to provide the requisite knowledge, technology, and decision-making methods to support participatory watershed management in southwest China. With the assistance of Oxfam-America, Green Watershed established—and now facilitates—the Lashi Watershed Management Committee. This committee runs dialogues among a broad range of government and community stakeholders to help them evaluate watershed development and protection options. In order to promote broader multi-stakeholder participation in the decision-making surrounding dams in southwest China, Green Watershed set up some exchanges bringing villagers from the Nujiang basin to visit to villages at the Manwan and Xiaowan Dams. This village-to-village visit enabled the Nujiang basin villagers to see first-hand the potential detrimental effects of dam building on remote rural communities. After the exchanges, Green Watershed gave these villagers an opportunity to voice their opinions to the public through the news media. Such reports allowed grassroots voices to be heard in the Nujiang dam decision-making.

Huai River Protectors (Henan Province)

Huo Daishan, the founder of this NGO (registered in 2003 at the bureau of civil affairs in Shenqiu county, Zhoukou city, in Henan province), is a photographer and journalist who has used photo exhibitions to help promote information on the severity of human health and ecological damage stemming from the extremely polluted Huai River. Mr. Huo also has conducted health surveys of villagers in the river basin and discovered abnormally high cancer rates, which appear to be caused by the water pollution in the rivers. His numerous health surveys along the polluted river and canals estimated over 100 villages have abnormally high rates of cancer patients. Huo also has used some funding from small private foundations to send water filtration equipment and medicine to some villages. CCTV and other news media organs have reported on his health surveys and assistance activities in these cancer villages. Such news reports have pushed local governments to invest into drilling deep wells to supply safe and clean water for villagers.

the Environmental Protection Control Act (1989) and WPPL, the Water Law authorizes MWR to oversee China's "water resource management," which has led MWR to view water quality protection as one of its responsibilities. Thus, there exists a contentious political struggle between SEPA and MWR over regulating water pollution.[15]

One major problem stemming from this struggle is that SEPA and MWR compete in collecting data on water quality and in regulating water pollution, which creates costly redundancies and severe lack of coordination in water clean up. Similar to the water quantity problems, the unwillingness of local governments to enforce water pollution laws and the weakness of the center to pressure them has led to a phenomenal growth in degraded watersheds.[16] The National People's Congress will finish revisions of the WPPL in early 2006, which will include provisions to strengthen enforcement and to clarify responsibility for regulating water pollution.

Obstacles to Integrated River Basin Management

While the 2002 reforms have on paper greatly increased the power of RBCs, these institutions need more reforms and capacity building to be fully capable of implementing integrated river basin management (IRBM). China's RBCs are merely extensions of the MWR and take a very top-down and narrow approach to manage the river basin. Besides lacking complete authority over water quality issues, China's RBCs do not have institutions to allow for local government or citizen input. In fact, the name river basin conservancy commission (*liuyu weiyuan hui*) is a misnomer, for China's RBCs have neither commissioners nor any official mechanism for provincial or local governments to shape polices or allocations in these top-down structures of river management.

Lacking a formal seat at the table has led provinces to undertake considerable backroom bargaining that ultimately hinders effective management of the rivers. For example, in 2002 the Yellow River flow barely made it to the ocean, which meant the furthest downstream province Shandong faced a severe water shortage that threatened the fall harvest. The Shandong government sent a delegation to Beijing that successfully lobbied for water to be released from reservoirs in Inner Mongolia, which caused water shortage and hardships upstream.[17] Such *ad hoc* decision-making to solve water problems ultimately causes more conflicts and does little to push forward water conservation or ecosystem protection.

New EIA Law

Over the past few years SEPA has been advocating increased public participation as another method for promoting better protection of natural resources and human health. The rights of the public to influence environmental policy formulation and implementation were vaguely guaranteed in China's first Environmental Protection Law in 1979. However, citizens' rights to influence environmental laws and infrastructure projects have only begun to be clarified and strengthened since the passage of recent legislation such as the 2003 Environmental Impact Assessment (EIA) Law. The first EIA Draft Law passed in the mid-1990s applied only to construction projects, but this new law requires evaluation of the plans for infrastructure and other construction. Notably, EIA reports must now be published and available for public comment.[18]

This new EIA Law already has empowered SEPA to exercise its muscle to protect the environment. For example, in January 2005 SEPA temporarily suspended 30 large construction projects across the country—many of which were dam and other water infrastructure projects that had neglected to develop proper EIA reports. Most of the suspended projects were quite large, including the Xiluodu hydropower plant along the Jinsha River in the upper reaches of the Yangtze River.[19]

Pan Yue, the outspoken Vice-Minister of SEPA, noted that this first "victory" in suspending particularly damaging factory and infrastructure projects does not mean SEPA has the capacity to comprehensively check all such projects. Pan stressed that public participation in the EIA process is also needed. He noted that SEPA intends to hold public hearings and forums so the public can get more involved in the EIA process.[20] Because SEPA and EPBs currently lack clear procedures to conduct outreach and hold such hearings, some international NGOs, such as the American Bar Association, have been holding trainings on how to carry out such hearings and other public outreach mechanisms. Moreover, the Japanese International Cooperation Agency (JICA) is assisting SEPA in drafting implementation guidelines for public participation in EIAs. The Chinese government is not only welcoming international groups to help in improving environmental protection regulations and projects, but also is increasingly open to domestic NGOs working conservation and environmental education on issues.

Bottom-Up Initiatives to Address China's Water Woes

The Missing Piece—Public Participation in Water Protection and River Basin Governance

On 13 November 2005, an explosion occurred at a PetroChina chemical plant in Jilin province that released over a hundred tons of benzene into the Songhua River. The Songhua flows into Heilongjiang province where it supplies drinking water for the provincial capital of Harbin and another 600 kilometers downstream it is the main water supply for the Russian city of Khabarovsk. For several days provincial and local officials in Jilin hesitated to inform downstream governments or SEPA about the spill. Once informed, Harbin officials also tried to cover up the crisis, first by telling city residents ten days after the spill that the water supply system would be cut off for "routine maintenance." However, in the face of growing rumors of a major chemical spill municipal officials quickly revised their announcement stating that the water system would be shut down for four days to prevent citizen exposure to benzene.[21] Many citizens fled the city, for few had confidence that the local officials could be trusted in providing accurate information on the health risks of the benzene.

The Chinese news media initially was quick and sharp in its criticism of the inadequate local response to the crisis, but toned down the negative reporting after a few days to highlight the efforts of the central government (which included investigations and disciplining of local officials).[22] The Minister of SEPA, Xie Zhenhua, was asked to resign in light of what was perceived as SEPA's initial inadequate response to the crisis. This case, exemplifies local protectionism of industry, shortcomings in emergency preparedness, and insufficient government transparency, as well as a pervasive lack of mechanisms for informing and involving the public in environmental protection issues.

Environmental NGOs Take the Stage

The move to increase the public's role in the environmental policymaking sphere in China began in 1994 under new administrative regulations that permitted the registration of "social organizations" (e.g., NGOs). The central leaders permitted this political opening to civil society because they knew the government needed broader help from citizens in addressing the growing social and environmental ills produced by rapid economic growth and the dismantling of the social welfare system. Admittedly, this political opening still has limits, for the registration regulations remain fairly restrictive in that they require all Chinese NGOs to obtain a government sponsor (referred to as *popo* or mother-in-law) and does not permit them to open branch offices. Another legal obstacle to limit the numbers of NGOs is the provision that no two groups can pursue the same kind of work within the same city or province.[23]

In 1994, the first environmental grassroots group to register under the new regulations was Friends of Nature, an environmental NGO. Other green groups also sought registration and those that failed often registered as businesses or operated without formal status. An increasing number of green groups have been established as solely Internet groups, thereby bypassing the registration system all together.[24] Today environmental NGOs number nearly 2,000 in China and have become the vanguard of civil society development. Initially, Chinese environmental NGOs tended to pursue "safe" activities such as promoting environmental education for schools and informing the general public on issues such as recycling, water conservation, and animal protection.

Despite registration challenges and the pressures to be non-confrontational, by the late 1990s a number of groups began increasing their area of operation—both geographically and thematically—which greatly enhanced their policy influence. While most Chinese green NGOs operate in urban areas or focus on biodiversity hotspots in Sichuan and Yunnan provinces, a handful of Chinese NGOs have been effectively working on water issues—particularly concerning river protection and public participation.

NGOs Diving into Water Work

While still few in number, some Chinese NGOs have been working on watershed and river protection initiatives, most of which include a strong public participation component. (See Box 3). One unique Chinese NGO with a broader focus—the Center for Legal Assistance for Pollution Victims (CLAPV)—has been playing a key role in helping victims of water pollution. Although the 1979 Environmental Protection Act (revised in 1989) granted pollution victims the right to sue in cases of damage, in practice it is challenging for citizens to navigate their way through the courts, which are often pressured by local governments to protect local industries. Over the past few years, private lawyers have been helping victims of major water pollution incidents win their cases. Like CLAPV, these lawyers are setting legal precedent and pushing the courts to build up their capacity in dealing with such cases, which often demands special expertise from the judges.[25]

The beginning of a dam on the first bend of the Yangtze (Jinsha section) is arousing considerable attention since it is being built on the main trunk of the river rather than tributaries. This dam will be the first of a planned 12-dam cascade on the entire Jinsha section, ending at the Tiger Leaping Gorge. These planned dams in the upper reaches of the Yangtze are prioritized not only to generate hydroelectric power, but also to prevent the Three Gorges Dam reservoir from becoming over-silted.   (Photo Credit Ma Jun)

A major watershed in the development of Chinese environmental NGOs took place in 2004 and 2005 when environmental activists and journalists built up a national campaign to push for more transparency in the construction of a series of 13 hydroelectric dams in Yunnan province on the Nu River (Nujiang)—one of two remaining wild rivers in China. In the fall of 2004, some Chinese environmental activists learned about the Yunnan provincial government's plans to construct these dams on the Nujiang, which led them to arrange for some journalists to tour the basin to investigate the dam plans and potential impact on the area.[26]

After the first group of journalists who traveled to Nujiang began reporting on the beauty of the area, which is notably a World Heritage Site, other journalists flocked to the basin. Within weeks hundreds of news stories and broadcasts across China were condemning the planned dams and the lack of transparency in their planning—they had not undergone the required environmental impact assessment (EIA). Environmental NGOs created a network organization called China Rivers Network to coordinate their joint work setting up photos exhibitions around the country to highlight the beauty of this endangered river to the public and to send petitions to central leaders.[27]

This extensive public debate caught the attention of Premier Wen Jiabao, who in February 2005 ordered the planning of the dams suspended pending an EIA. In August 2005, a broad coalition of Chinese groups (which included 61 NGOs and 99 researchers and government officials) sent an open letter to the top leaders urging public disclosure of the EIA for the hydropower development plan on the Nujiang before the government approves any dams on this free–flowing river.[28]

As this report was completed, the debate on the dams was still ongoing. Even if the dams do begin construction the campaign represents a major victory for Chinese environmentalists, who in partnership with journalists brought this issue into an open debate. The campaign, which was built on a decade of steady development of NGOs working with (or generally not against) the Chinese government, is also a testament to the increased freedoms Chinese environmentalists have come to enjoy. It should be stressed that NGO activities around the Nujiang are not "anti-dam" campaigns, rather a broader push for more transparency and citizen participation in water management and environmental policymaking in China.

Bringing Together Top-Down and Bottom-Up Water Work

All nations struggle with implementing water protection laws, but China's obstacles are particularly challenging—population pressures, rapid economic growth, bureaucratic infighting, unclear water rights, and local government protectionism. The Chinese government has established a strong foundation of laws and regulations to prevent water pollution and strengthen water conservation laws. Moreover, the government has forged partnerships with multilateral organizations and international NGOs to help address water management and river protection challenges.

Besides inviting international expertise, the Chinese leadership has permitted considerable political space for Chinese environmental NGOs to expand, for they know the government cannot solve environmental (particularly water) problems solely from the top-down. Over the past few years SEPA officials have been emphasizing the need to increase the public's role in shaping environmental laws and monitoring local governments and industry, for such bottom-up participation ultimately will decrease the government's regulatory and fiscal burden in enforcing environmental regulations. In July 2004, the State Council passed the Administration Permission Law (1 July 2004), which requires administrative agencies to inform citizens that they have a right to express their opinions at a hearing regarding any government project that impacts them.[29] SEPA was notably the first agency in China to write regulations and actually hold public hearings based on this new law.

Another sign of increasing transparency in the environmental sphere occurred in the fall of 2005 when SEPA circulated for comment a draft regulation that aimed to improve public participation in the EIA process. This new regulation contains stipulations on protecting participants' rights, disclosing information, and designing new procedures for public involvement. This solicitation for comments represents the first time any Chinese government agency has openly called for public input on a new regulation.[30]

A similar push for more openness in environmental information occurred in November 2005 when SEPA made a call for nationwide implementation of corporate environmental performance rating and disclosure.[31] In December 2005, the State Council included a provision in its *Decision on Environmental Protection* that requires industries to "publicly disclose their environmental information."[32]

Against this backdrop of positive trends of openness towards public disclosure of information, NGO development, and public participation in the environmental sphere, the government has become somewhat concerned of what is perceived as too much social activism in China. This wariness stems from the growing number of protests throughout China on a whole range of issues.[33] Local governments have been particularly sensitive to NGOs monitoring polluting factories.[34] However, some municipal governments such as Shenzhen, Beijing, and Xiamen have been very welcoming of NGO and citizen involvement in the environmental sphere. More international projects to help stress the utility of public participation in helping the government reach its water protection goals could help mitigate concerns about social activism. While green NGOs in China face some external constraints, they also are hampered by some internal shortcomings that threaten their sustainability in the long run—overdependence on foreign assistance, lack of internal transparency, and high turnover in staff due to low paying positions.

In order for the Chinese government to move forward in strengthening water pollution and management laws, it will need not only to continue reforming the river basin commissions and laws from the top down, but also pushing reforms to promote stronger environmental NGO development and citizen participation. Some needed reforms include: (1) revising the rules to make registration more accessible to NGOs; (2) pushing forward tax-free donation regulations to encourage Chinese businesses and citizens to give to local NGOs, thereby breaking NGO dependence on international organizations; and (3) significantly increasing the access of the public and NGOs to information of environmental decision making (e.g., new EIA Law) and project implementation.

Notes

1. Baldinger, Pamela and Jennifer L. Turner. (2002). *Crouching Suspicions, Hidden Potential: U.S. Energy and Environmental Cooperation with China.* Washington, DC: Woodrow Wilson Center.

2. "China Politics: Green-tinted Glasses." (2004, July 6). *The Economist Intelligence Unit.*

3. Yin Runsheng, Jintao Xu, Zhou Li, and Can Liu. (2005). "China's Ecological Rehabilitation: The Unprecedented Efforts and Dramatic Impacts of Reforestation and Slope Protection in Western China." *China Environment Series*, Issue 7, 17-32.

4. Chinanews.cn. (2005, November 25). "China's Water Quality is Improving." [Online]. Available: http://www.chinanews.cn//news/2005/2005-11-25/14799.html

5. *Shenghuou Zhoukan* [*Life Weekly*]. (2005). Volume 10, 27.

6. The complete Eleventh FYP can be found in Chinese on the Xinhua News Agency Web site: http://news.xinhuanet.com/politics/2005-10/18/content_3640318.htm

7. U.S. Embassy in Beijing. (2005, November). *Beijing Environment, Science and Technology Update.*

8. Xie Zhenhua. (2005, November 8). *Talk with SEPA Minister.* Presentation at the Woodrow Wilson Center's China Environment Forum, Washington, DC.

9. Wang Yahua. (2005)."River Governance Structure in China: A Study of Water Quantity/ Quality Management Regimes." In Jennifer L. Turner and Kenji Otsuka (Eds.), *Promoting Sustainable River Basin Governance: Crafting Japan-U.S. Water Partnerships in China.* IDE Spot Survey No. 28. (pp. 23-36). Chiba, Japan: Institute of Developing Economies/IDE-Jetro. [Online]. Available: http://www.ide.go.jp/English/Publish/Spot/28.html

10. Peisert, Christoph and Eva Sternfeld. (2005). "Quenching Beijing's Thirst: The Need for Integrated Management for the Endangered Miyun Reservoir." *China Environment Series*, Issue 7, 33-45.

11. Wang Yahua. (2005).

12. Ibid.

13. Economy, Elizabeth. (2004). *The River Runs Black: The Environmental Challenge to China's Future.*

New York: Cornell University Press.

14. Wang Yahua. (2005).

15. Ibid.

16. Hildebrandt, Timothy and Jennifer L. Turner. (2005). "Water Conflict Resolution in China. *China Environment Series*, Issue 7, 99-103.

17. Wang Yahua. (2003). "Water Dispute in the Yellow River Basin: Challenges to a Centralized System." *China Environment Series,* Issue 6, 94-98.

18. Tang, Shui Yan with C.P.Tang and Carlos Wing-Hung Lo. (2005). "Public Participation and Environmental Impact Assessment in Mainland China and Taiwan: Political Foundation of Environmental Management." *Journal of Development Studies*, Vol. 41, No. 1, 1-32.

19. Qin Chuan. (2005, February 3). "All 30 Law-breaking Projects Suspended." *China Daily,* p. 2.

20. Jason Subler. (2005, February 9). "China Crackdown on Assessment Violations Could Reflect Long-Term Enforcement Trend." *International Environment Reporter,* Volume 28 (3). [Online]. Available: http://ehscenter.bna.com/pic2/ehs.nsf/id/BNAP-69FH3P?OpenDocument

21. Luis Ramirez. (2005, 23 November). "Residents Flee Chinese City as Taps Go Dry Over Water Poisoning Scare." Voice of America. [Online]. Available: http://www.voanews.com/english/2005-
11-23-voa23.cfm

22. Xinhuanet. (2005, November 23). "River Pollution Spurs Measures for Cleaner Water." *China View.* [Online]. Available: http://news.xinhuanet.com/english/2005-11/25/content_3835575.htm

23. Nick Young. (2001). "Searching for Civil Society." In *Civil Society in the Making: 250 Chinese NGOs* (pp. 9-19). Beijing: China Development Brief.

24. Guobin Yang. (2003). "Weaving a Green Web: The Internet and Environmental Activism in China." *China Environment Series*, Issue 6, 89-93.

25. Ma Jun. (2003). "Sue Me Sue You Blues." *China Environment Series*, Issue 6, 81-83.

26. Hu Kanping with Yu Xiaogang, "Bridge Over Troubled Waters: The Role of the News Media in Promoting Public Participation in River Basin management and Environmental Protection in China." In Jennifer Turner and Kenji Otsuka (Eds.), op. cit. (pp. 125-140).

27. Jim Yardley. (2004, March 10). "Dam Building Threatens China's Grand Canyon," *The New York Times*, p. A1.

28. Translation of the NGO letter available on the International Rivers Network Web site: http://www.irn.org/programs/nujiang/index.php?id=050903disclose_pr.html

29. Ibid.

30. Yingling Liu. (2005). "China to Strengthen Public Participation in Environmental Impact Assessments." *China Watch*. [Online]. Available: www.worldwatch.org/features/chinawatch/stories/20051209-1

31.SEPA. (2005, November 21). *Opinion on Speeding up Evaluation of Enterprise Environmental Behavior.* [Online]. Available: http://www.sepa.gov.cn/eic/649086798147878912/20051124/13170.shtml

32. China State Council. (2005, December 3). *State Council Decision on Environmental Protection.* [Online]. Available: http://www.zhb.gov.cn/eic/649096689457561600/20051214/13756.shtml

33. Cody, Edward. (2005, November 26). "In Chinese Uprisings, Peasants Find New Allies: Protesters Gain Help of Veteran Activists." *The Washington Post*, pp. A1, A16.

34. Buckley, Lila and Jennifer L. Turner. (2005, November 23). "Environmental Activist Arrested in Hangzhou; Movement Still Hampered by Legal and Financial Restrictions." *China Watch.* [Online]. Available: www.worldwatch.org/features/chinawatch/stories/20051123-1

PART THREE:
FLOW OF INTERNATIONAL AID
TO PROTECT CHINA'S RIVERS

Over the past twenty years, many international organizations have worked with the State Council, National People's Congress, State Environmental Protection Administration (SEPA), MWR, and other ministries to develop new environmental policies, regulations, and pilot projects. The increasing health threats and conflicts arising from water pollution and scarcity in the 1990s led the Chinese government to request more international assistance in this sector. Below we provide an overview of international projects in the area of water and river protection in China. Although international initiatives in this sector are growing, there exist many more opportunities for such work in China—an issue highlighted in Part IV.

International River Basin Initiatives in China

Multilateral Aid

World Bank. China is the World Bank's largest recipient of loans and grants in the environmental sphere across a broad range of sectors—air pollution control, grassland protection, information disclosure, and water resource protection. The World Bank has been involved in numerous water protection projects including two notable initiatives aimed directly at improving the capacity of river basin governance institutions in the Tarim and Hai river basins.

Tarim River Basin. In Xinjiang, the World Bank undertook a challenging project to create a new river basin commission for the Tarim River. This project has established China's first truly "participatory" river basin management commission. Although it would be difficult to translate this experience to a larger basin with many more stakeholders, it could be useful for other international organizations to do similar projects on other small river basins to build up more support for such institutions within the Chinese government sphere.

Hai River Basin. In 2004, with $17 million in Global Environment Facility (GEF) grant money, the World Bank began a project on the Hai River Basin that aims to accelerate the integration of water and environmental management in the basin. The main challenge of this project is bringing together SEPA and MWR to jointly undertake the institutional reforms necessary to establish mechanisms for local water and environment bureaus to truly work together. The project also aims to improve the technologies to undertake integrated water planning.

Asian Development Bank.[1] Since 1986, China has been the second largest member of the Asian Development Bank (ADB) and one of its best performing portfolios. The environmental projects supported by ADB cover a broad range of issues—energy efficiency development (including renewables), urban environmental protection, and water management reforms. Water-focused initiatives have had a strong focus on municipal water management (both wastewater and water supply), as well as wetland protection (Sanjiang Plain) and river basin pollution control (Hai River). Beginning in 2003 ADB initiated a major study on the Yellow River titled the Trans-jurisdictional Environmental Management project. This study is cross-sectoral focusing on legal, financing, management, and social challenges to protect the Yellow River. The first phase of the project examined water management laws and practices at both the national and local levels. The study also evaluated mechanisms for intergovernmental relations in the basin, particularly surrounding trans-jurisdictional water pollution conflicts. This first phase offered recommendations to the Chinese government on revising current laws, as well as creating new legislation and cooperative mechanisms (e.g., a joint-ministerial committee) to promote better coordination between agencies and monitoring systems on the basin. The second phase of this project includes analysis on improving the financing of river protection focusing on the Wei River, a tributary of the Yellow River. This financing study aims to prepare specific recommendations for China's State Council on the management and financing of water control projects. Many existing water control laws are too broad and general. Therefore, this extensive ADB research initiative aims to provide a detailed technical background to better prepare MWR, SEPA, and other central agencies to reform laws to better protect China's rivers.[2]

Bilateral Aid

The United Kingdom's Department for International Development. The UK Department for International Development (DFID) focuses on poverty reduction through partnerships with developing country governments. In China, DFID's water work focuses on improved livelihoods and health through better water management and sustainable access to safe water and sanitation. DFID, in partnership with other international agencies, is supporting the Chinese government in implementing programs supporting water sector reforms proposed in the 2003 revision of the Chinese Water Law, including: increased user participation, more integrated approaches to water resource management, new approaches to soil and water conservation, and increased access to drinking water and sanitation.

The European Union.[3] Between 2002-2006 the European Union (EU) budgeted EUR 250 million for collaborative projects with China, 30 percent of which was devoted to initiatives to promote environmental protection and sustainable development.[4] One of the largest environmental projects focuses on the Liao River Basin in Liaoning province. Poor regulation of heavy industries and agricultural runoff has made the Liao one of the most polluted rivers in China. Over-extraction of Liao water has created a severe water shortage that has left Liaoning province with 603 cubic meters (m^3) of water per person compared with the national average of 2,292 m^3. For over five years the EU office in Beijing—with support from the EU, Japan, and the World Bank—has been working with the Liaoning provincial government to create and implement a broad range of projects promoting sustainable river basin management in this highly stressed basin. The EU and their Chinese partners are working to create an integrated framework for pollution control and water resource management by: (1) setting up pilot catchment water quality protection plans for one of the main reservoirs (Dahuofang Reservoir); (2) undertaking investigations of industrial water conservation and pollution; and (3) developing water quality models for the entire basin using GIS and decision analysis software. These project activities already have enabled the EU team to make basin-wide recommendations on reforming water sector institutions and tariffs, which were adopted into the Tenth Five-Year Plan of Liaoning province.

Swedish International Development Cooperation Agency. Over the past several years Sida—the Swedish International Development Cooperation Agency—has been supporting a number of water-related cooperation projects in the following areas: (1) more water efficient technologies in industry, (2) preparation of a comprehensive action plan for the restoration of a lake in Inner Mongolia, (3) water efficiency in agriculture, (4) capacity building for better management of wastewater treatment plants, and (5) development of an ecologic sanitation system. Sida is one of the main financiers of the China Council for International Cooperation on Environment and Development (see below). Within the council water resource management is one key work area. Sida also provides concessionary credits for construction of wastewater treatment plants in China. Sweden's new cooperation strategy for China (2006-2010) will continue to prioritize environmentally sustainable development.

Switzerland's Bilateral Aid. In 2005, most of Switzerland's environmental bilateral aid focused on research and projects in Sichuan province. In 2005, Swiss researchers and Chinese counterparts conducted research, training, and demonstration projects on ecotourism, which included protective tourism development of World Heritage sites and Shangri-La Mountain, as well as eco-agricultural tourism on Longquan and Sancha lakes. In 2006 Switzerland's work in Sichuan province will include studies on water and environmental management in the Min and Tuo river drainage areas. These studies will examine management and policy options for preventing water pollution from the industrial, urban, and rural sectors in the drainage area. Another water-related study will be conducted in 2007 on the construction of ecological defense in the upper reaches of the Yangtze River. These studies will focus on restoration of forest ecosystems through sustainable ecological forestry in the upper basin, which will help prevent erosion and serious floods in the Yangtze.

China Council for International Cooperation on Environment and Development.[5] The China Council for International Cooperation on Environment and Development (CCICED) is a high-level consultative body made up of both international and Chinese experts and officials who provide strategic consultation to China's State Council con-

Instead of properly processing wastewater a paper mill in the middle reaches of the Ganjiang built a fence to catch foamy emissions so they can be diluted before being released into the river. (Photo Credit: Xiao Qiping)

cerning environment and development issues. In March 2003, the council launched a task force on integrated river basin management (IRBM), focusing on the Yellow River. The overall objective of this task force, which includes considerable NGO participation, was to promote healthy river basins in China through better governance of water resources, biodiversity conservation, and ecosystem management through information sharing, demonstration, and public participation. In addition to gathering information on how IRBM is implemented around the world, the IRBM task force worked with WWF-China on studies in the Yangtze River to create a basin conservation plan.[6] The ideas in this plan have been shared with local and central government agencies, as well as community groups to solicit input for the final version they will present to CCICED with the hope of informing future legislation and pilot projects.[7]

Japanese Government's Water Work in China

Since the mid-1990s, a very significant share of all Japanese ODA for China has been for environmental projects. A large number of these projects focus on water, particularly wastewater treatment, water supply facilities, water conservation in large-scale irrigation districts, and river basin improvement. In 2004, the Japanese government announced it will prioritize water resource management and conservation in China through afforestation, anti-desertification, and watershed management. Moreover, Japan will build on previous bilateral assistance in China to address problems of water pollution and ecosystem conservation. The Japanese government's ODA contribution to environmental protection in China has been predominantly Yen loans for infrastructure. However, the Japanese government agencies are discussing with their Chinese counterparts the possibility of terminating Yen loan projects to China by 2008, which coincides with the year

Beijing is hosting the Olympics. If Yen loans are terminated, Japanese ODA to China is expected to focus more on assistance to institutional reform and human resource development issues, rather than infrastructure projects.

Japan Bank for International Cooperation. In China, Japan Bank for International Cooperation (JBIC), which supplies Yen loans to many developing countries, focuses on three target areas: environment, human resource development, and poverty alleviation in the western region. Since 1979, JBIC (previously called OECF) has made significant loan commitments to China, over the last five years JBIC loans have averaged $1.5 billion a year.[8] JBIC does not have specific projects to support river basin management, but is involved in many water-related projects such as: (1) water supply projects in more than 20 large cities in China; (2) water pollution control projects supporting industrial wastewater treatment and sewage plant construction and their expansion in rivers basins in five provinces—Huai River (Henan), Songhua-Liao River (Jilin), Songhua River (Heilongjiang), Xiangjiang River (Hunan), and the upper Sanxia Dam River (Chongqing City); (3) water saving irrigation in Xinjiang and Gansu; (4) afforestation on the Loess Plateau (in Shaanxi, Shanxi, and Inner Mongolia), in which one central goal is to greatly reduce siltation of the Yellow River; (5) afforestation in Hubei and Jiangxi provinces in the middle part of Yangtze River; and (6) multipurpose dams for flood control and water supply in Sichuan, Henan, and other provinces.

Japan International Cooperation Agency. Japan International Cooperation Agency (JICA) projects related to water include a technical cooperation project in which Japanese experts are dispatched to train counterparts from China or Chinese experts are invited to Japan for training in areas such as: (1) human resource development for a water resources project, in which JICA aims to train more than 2,000 central and local government water bureau personnel; (2) model planning project for water-saving measures in large-scale irrigation schemes; (3) a water environment restoration pilot project in Lake Tai; and (4) a model planning of afforestation in Sichuan province. In Xinjiang, JICA is undertaking a development study of sustainable underground water in the Tulufan Basin and a development study of comprehensive landslide disaster control in Yunnan province in the Xiaojiang River, a tributary of upper Yangtze River.[9] Moreover, JICA is working with the Chinese Ministry of Construction, MWR, as well as local and provincial governments to develop an instruction manual for promoting water saving in irrigation.

While most of Japan's water assistance has focused on technology transfer—the "hardware" of water management—in the last several years, JICA projects have included some initiatives aimed at strengthening China's human capital and policies in the water sector—the much-need water management "software." Moreover, JICA has just started to set up a water rights project in China drawing on assistance from scholars affiliated with the Japanese Ministry of Land and Transportation and Japanese universities, and conduct a case study of the Taize River in Liaoning province. As part of a JICA model planning project to promote water saving measures in large-scale irrigation schemes, the Japanese Institute of Irrigation and Drainage has been conducting a technical information exchange program in China with the goal of introducing the Japanese experiences

BOX 4:
U.S. NGOS AND UNIVERSITIES INVOLVED IN WATER WORK IN CHINA

WWF-China. WWF-China has several major integrated river basin management initiatives on the Yangtze River, which include demonstration projects to improve flood control by restoring wetlands and lakes and increasing public participation in water management through community education and NGO capacity building activities. In 2005, WWF established a small grants program that funded 22 projects aimed at promoting the conservation of Yangtze aquatic species.

Conservation International. Since 2005, Conservation International (CI) has been working with The Nature Conservancy and China's State Forestry Administration on developing payment for environmental services (PES) system for carbon and water in southwest China. A pilot project in Lijiang, Yunnan Province (part of CI's Forest for Climate Community and Biodiversity project) is being planned to work on watershed protection and reforestation issues with upstream farmers and downstream water users in the city of Lijiang. CI also is collaborating with the Environment and Natural Resource Protection Committee of China's National People's Congress in research and projects to help inform the creation of PES legislation in China.

Roots & Shoots. Roots & Shoots is a project of the Jane Goodall Institute (JGI) China. In early 2006 JGI China will begin cooperation with the Chengdu Urban Rivers Association (CURA, a Chinese NGO) on a "model eco-village" project in rural Sichuan as part of a watershed clean-up project. In an attempt to cleanup the water sources supplying the city, CURA is working in watersheds upstream to address issues of agricultural run-off from chemical fertilizers and pesticides. CURA, Sichuan University, JGI China, and Roots & Shoots, will be working with one upstream village to take an integrated approach to addressing runoff problems by working simultaneously on environmental education, organic agriculture, and local livelihood issues.

The Nature Conservancy. In partnership with the government of China and related academic institutions, The Nature Conservancy (TNC) is developing a comprehensive, scientific map of the distribution, representation, and viability of China's important biodiversity. Within this initiative the National Development and Reform Commission and SEPA are for the first time working in collaboration on a single plan to inform sustainable economic decision-making, and to redesign and expand China's protected area system. As part of this partnership, TNC will develop extensive databases for assessing and monitoring freshwater biodiversity in the upper Yangtze region and establish conservation priorities and strategies for protecting those resources in this important region that is home to approximately 350 million people. TNC has also catalyzed an assessment of sustainable energy options for an integrated power grid in which hydropower development is designed to the greatest extent possible to conserve freshwater ecosystems and sustain local livelihoods.

BOX 5:
JAPANESE RESEARCH CENTERS AND NGOS INVOLVED IN WATER WORK IN CHINA

Ramsar Center Japan. Ramsar Center Japan (RCJ) has been actively involved in research and public awareness of wetlands in Japan, China, and Asian countries. In China, Ramsar Center Japan (RCJ) has collaborated with the Beijing-based office of Wetlands International-China to conduct environmental education and exchange programs (held in Dafeng, Jiangsu province in 2004 and Zalong nature reserve, Heilongjiang province in 2005) on wetland preservation for primary and junior high school children from China, Korean and Japan.[15]

Japan Environmental Council. In November 2005, some members of the Japan Environmental Council (JEC) visited Henan province to investigate opportunities for cooperation with a unique local NGO the Huai River Protectors (see Part II, Box 2) on water pollution problems in the Huai River. One member of JEC is planning to invite the NGO's founder Huo Daishan to Minamata in September 2006 to hold a photo exhibition for the Japanese public to help boost exchanges with other Asian NGOs that work with pollution victims.

Mekong Watch. Mekong Watch is a Tokyo-based watchdog and policy research NGO for the Mekong River.[16] In 2005, Mekong Watch dispatched one staff to Kunming, Yunnan province to conduct research, jointly with a local NGO Green Watershed (See Part II, Box 2), about potential threats dam building and other development in the upper reaches of the Mekong River pose to the environment and citizens living in the basin in Yunnan province, as well as downriver.

Japan-China New Century Association. Japan-China New Century Association con-ducted a Japan-China Water Forum in Beijing (April 2004) and in Sapporo (October 2005) with its counterpart China National Youth Union to boost exchanges between governmental officials, scholars, business representatives, and NGO activists in both countries around the issue of water resource protection.

Japan-China Water Forum. Since 2004, the Japan Water Forum[17] has carried out a Japan-China Water Meeting to invite water experts in Japan and China to meet and exchange information to promote mutual understanding on water issues between both countries.

including Land Improvement Districts (LID). LIDs have been used in Japan for over forty years and represent a successful participatory irrigation management practice highly relevant for China.[10]

U.S. Government Water Work in China

In stark contrast to Japan, U.S. government agencies do not provide loans or grants to the Chinese government for environmental projects. Due to congressional restrictions on direct assistance to China, the nearly 20 U.S. government agencies currently carrying out over 100 environmental or energy initiatives in China support their work by internal agency budgets not formal development assistance.[11] Despite limited funding, under the 1979 U.S.-China Scientific and Technology Cooperative Agreement the two countries have signed thirty protocols which form the foundation for joint projects, research, and information exchange on natural resource protection, atmosphere, marine health, pollution and energy issues. While air quality and energy efficiency related projects are the key areas of cooperation, research, and exchange between the United States and China, over the past few years U.S. agencies have undertaken some comprehensive water projects in China. Some examples of U.S.-government led water initiatives are outlined below.[12]

Department of Agriculture and Environmental Protection Agency. Since 2000, U.S. Department of Agriculture (USDA) and U.S. Environmental Protection Agency (EPA) have been conducting water quality monitoring, wastewater reuse, and watershed management demonstration projects on the lower reaches of the Yellow River. For example, since 2001, EPA and USDA in collaboration with MWR and two provincial environmental protection bureaus (EPBs) have been conducting joint wastewater treatment and monitoring demonstration projects in Shandong and Henan provinces.

USDA's Economic Research Service has been working with the Chinese Academy of Sciences, MWR, Australian Bureau of Agricultural and Resource Economics, and University of California, Davis in conducting research into water resource and agricultural production issues in China. The core of their cooperation since 2003 has been collecting data for a Yellow River Basin Model, which in 2005 produced preliminary scenarios of water trades and environmental flows in the basin. The scenarios indicate that there are substantial gains to be made from water trades that could actually increase grain production. These partners are also conducting surveys in the basin analyzing water-saving technology adoption and the creation of water user associations and canal contracting reforms for irrigation districts.

In 2006, USDA's Economic Research Service is proposing a new initiative with researchers at MWR to better understand the hydrological implications of irrigation management reform and water-saving irrigation technology adoption so they can incorporate the effects of the impacts into the hydrological component of the Yellow River Basin Model. In addition, they would focus research on the de facto property rights to water in rural China in order to propose formal property right systems that closely match current practices.

In 2006, EPA will complete a Clean Water for Sustainable Cities in China Program in the Hai River Basin. This water quality-focused project is being done

Fishing Boats near the headwaters of the Ganjiang, where the water is still relatively clean. (Photo Credit: Xiao Qiping)

in collaboration with the Tianjin Environmental Protection Bureau, SEPA, MWR, Hai River Conservancy Commission, Global Environment Facility (GEF), and ADB. This project aims to increase public access to safe drinking water and sanitation, and to promote watershed management in the Hai River Basin near Tianjin. The project is focusing on protecting the quality of source water at the Yuqiao Reservoir through improved management of waste and runoff from villages, hotels, restaurants, fishponds, and agriculture surrounding the reservoir. The project will advance the development of a watershed management plan in collaboration with the GEF Hai Basin Integrated Water and Environment Management Project. A new water pollution prevention initiative is emerging from a 2003 bilateral agreement signed between SEPA and EPA, which includes a Memorandum of Understanding to undertake pilot projects on water pollution trading in China.

International Environmental NGO and Research Institutes Working on Water in China

Over the past few years, international NGOs have begun to do more work in the area of river basin protection and management. Although some U.S. environmental NGOs are particularly active in Chinese river basin protection as a core or secondary area of work (see Box 4) Japanese environmental NGOs and research centers also have become quite active conducting study tours, joint studies, conferences, and workshops around the theme of river basins in China. (See Box 5). These international NGO and research institute water projects have been building networks that bring together (often for the first time) central, provincial, and local government agencies, research centers, and Chinese NGOs. In short, such projects are creating new lines of communication and increasing stakeholder participation around water protection in China.

Conclusion

The magnitude of the water problems in China and the government's openness to outside assistance has led to this growing involvement of international organizations in water management and pollution control throughout the country. Over the past few years, international organizations have moved from small project-oriented initiatives (e.g., wastewater treatment installation) to more ambitious basin-wide or national policy focused initiatives (e.g., reforms of RBCs and working on water rights issues). The number and scope of international water projects in China have grown, yet the organizations conducting this work are rarely sharing their project's successes or lessons learned. The variety of international river initiatives introduced above offers insights into potential options for U.S.-Japan collaboration on river basin governance in China, which we discuss in Part IV.

Notes

1. Information in this section is drawn from talk by World Bank senior irrigation specialist Liping Jiang to the China Environment Forum (CEF) and Institute of Developing Economies (IDE) study group in Beijing on 17 June 2004.

2. Information in this section drawn from talk by People's University Professor Ma Zhong and his research team to the CEF/IDE water study group at Tsinghua University in Beijing on 16 June 2004.

3. Hildebrandt, Timothy and Jennifer L. Turner. (2005). "Water Conflict Resolution in China." *China Environment Series*, Issue 7. (pp. 99-103).

4. Information in this section is drawn from talk by European Union representatives Alan Edwards and Wang Yongli to CEF/IDE study group at Tsinghua University in Beijing on 16 June 2004.

5. European Union. (2002). *China: Country Strategy Paper 2002-2006.* [Online]. Available: http://europa.eu.int/comm/external_relations/china/csp/index.htm

6. Information in this section is drawn from talk by CCICED IRBM Task Force representatives Yu Xiubo & Li Lifeng to CEF/IDE study group at Tsinghua University in Beijing on 16 June 2004.

7. IRBM case studies WWF has carried out in the Yangtze Basin include: (1) Xianghexi River Basin; (2) Lake Zhangdu River Basin on wetland and river basin management; (3) Minshan Mountain System to draw lessons from a landscape restoration project; (4) Lake Poyang where WWF has been working with local stakeholders (government, NGOs, and community groups) to devise an IRBM Action Plan; and (5) Danjiangkou Reservoir (upper Han River).

8. The CCICED report in English [Online]. Available: www.harbour.sfu.ca/dlam

9. Information in this section is drawn from presentation by JBIC representative Naoki Mori to CEF/IDE study group at Tsinghua University in Beijing on 16 June 2004.

10. Information in this section is drawn from talk by JICA representative Mr. Satoshi Nakamura to CEF/IDE water study group at Tsinghua University in Beijing on 16 June 2004.

11. Yamada, Nanae. (2005). "Irrigation and River Basin Management in Japan: Toward Sustainable Water Use." In Turner and Otsuka (Eds.), *Promoting Sustainable River Basin Governance: Crafting Japan-U.S. Water Partnerships in China. IDE Spot Survey No. 28* (pp. 83-101). Chiba, Japan: Institute of Developing Economies/IDE-Jetro. [Online]. Available: http://www.ide.go.jp/English/ Publish/Spot/28.html. Yamada, Nanae. (2005) "Situation and Tasks of Participatory Irrigation Management in China." *Ajiken World Trend*, Vol.122, 14-17. [In Japanese]. For more information on Japanese Institute of Irrigation and Drainage: http://www.jiid.or.jp/e/index.html

12. Turner, Jennifer. (2002 2003, & 2005). "Inventory of Environmental and Energy Projects in China." *China Environment Series*, Issues 5, 6, and 7. Washington, DC: Woodrow Wilson Center.

13. Ibid.

14. Ibid.

15. See RCJ web site at: http://homepage1.nifty.com/rcj/english/menu-top.english.html

16. See the web site at: http://www.mekongwatch.org/english/index.html

17. See the web site at: http://www.waterforum.jp/eng/

PART FOUR:

OPPORTUNITIES FOR JAPAN-U.S. COLLABORATION ON RIVER BASIN GOVERNANCE IN CHINA

Over the past fifteen years, despite the Chinese government's promulgation of ever-stronger water protection policies and more ambitious targets and campaigns to clean up major rivers and lakes, the quality of China's waters—particularly rivers—has decreased markedly. Domestic reforms of water laws and international assistance have helped push forward the concept of integrated water resources management in China. Nevertheless, the policy changes and international assistance necessary to mitigate China's water problems, particularly to protect river basins, will demand creative thinking and dialogue with environmental experts and practitioners from global, regional, national, and sub-national organizations. Assisting China on a path to sustainable river basin development in the coming decade is of such great importance that Japan and the United States in particular should do their utmost to cooperate—or at least to coordinate their efforts—in this important arena, and to share the benefits of their experience and technology.

The U.S. and Japanese governments (as well as NGOs and research institutes) are quite active in assistance and research on environmental protection issues (particularly water) in China. However, little information is shared and there are no formal joint initiatives between the U.S. and Japan. Economic slowdowns, shifting geopolitical priorities, and recent major natural disasters have led both the United States and Japan to make some cutbacks on overseas development assistance. Therefore, information sharing and joint work in international environmental assistance could enable both countries to increase the impact of their shrinking aid budgets as well as avoid investing in redundant projects in China and other developing countries.

Below we first discuss how the Japanese and U.S. governments have come to prioritize water in their international assistance programs. We then highlight some possible areas of collaboration between governments, NGOs, and research centers in the United States and Japan on protecting rivers in China along the three themes crucial for integrated river basin management (IRBM): (1) river basin management institutions, (2) financing mechanisms, and (3) public participation.

Water Priorities in International Assistance

Both the United States and Japan are giving water sector issues a high priority in their international assistance programs, often as part of broader poverty relief or urban development efforts in developing countries. In 2003 one major recommendation emerging from the Third World Water Forum held in Japan was the need for greater international cooperation on water issues in developing countries. In the spirit of this recommendation, the

Japan Water Agency and the Asian Development Bank initiated the Network of Asian River Basin Organizations (NARBO) project at the World Water Forum. Drawing on Japanese experience in water development and conservation, NARBO aims to promote integrated water resource management (IWRM) in river basins across Asia through advocacy, training, technical advice, and regional cooperation.

Within the U.S. Agency for International Development (USAID) the protection and environmentally sound development of the world's water resources is a top priority. In countries around the world, USAID projects and investments in the water sector have focused on improving access to safe and adequate water supply and sanitation, improving irrigation technology, protecting aquatic ecosystems, and strengthening institutional capacity for water resources management. Between 2003 and 2005, USAID invested over $1.7 billion to improve sustainable management of freshwater and coastal resources in more than 76 developing countries. During this same period more than 24 million people received improved access to freshwater, nearly 28 million people received improved access to sanitation, and some 3,400 watershed governance groups were convened to undertake basin-scale, integrated water resource management decision-making.[1] USAID may expand its water work in light of the Senator Paul Simon Water for the Poor Act of 2005 that was signed into law on 30 November 2005. This new act aims to make safe and affordable drinking water and sanitation, and sustainable water resources management a cornerstone of U.S. foreign policy.

In addition to independent assistance around the world, the United States and Japan are exploring ways to strengthen their water programs through cooperative efforts. At the World Summit on Sustainable Development in 2002, the U.S. and Japanese governments launched a new cooperative initiative on water (the U.S.-Japan Water Partnership), in which the two countries agreed to pursue joint or parallel water projects in developing countries. USAID and the Japan Bank for International Cooperation (JBIC) are now leading efforts to implement water financing projects and programs in four countries—the Philippines, Indonesia, Jamaica, and India. In the Philippines, for example, two pilot projects are well underway. In one project the Municipal Water Loan Financing Facility will tap a JBIC-supported credit facility and private investment that is backed by USAID's Development Credit Authority. Meanwhile, a feasibility study was completed in early 2005 for a new Philippine Water Revolving Fund, scheduled for launch in early 2007. Similar financing and related programs are under design in the other three pilot countries. While China is not currently targeted under this collaborative program, it is clearly a country that could benefit from such joint U.S.-Japan assistance on water financing.

Possible Areas of Collaboration

China's water needs are vast and complicated, however using the IRBM lens with a focus on management institutions, financing, and public participation we present a kind of "buffet" of ideas that are meant to help catalyze some thinking in the United States and Japan (as well as other countries) on potential areas of water collaboration in China. The potential for joint assistance lies not only between the Japanese and U.S. governments, but also between nongovernmental and research sectors in both countries.

Legislative and Institutional Reforms

To promote legal and institutional reforms to push the IRBM concept in China, joint work by the U.S. and Japanese governments could focus on setting up a small pilot program in one watershed (e.g., tributary, lake, or estuary) within one of the seven major large river basins, all of which have river basin commissions (RBCs). Pilot programs could focus on one small-scale institutional change—such as water rights, water user associations, or pricing. An even more ambitious pilot project, which was recommended by the CCICED (China Council for International Cooperation on Environment and Development) IRBM Task Force, would be the creation of tributary-, lake-, or estuary-level management commissions comprised of provincial governments, local administrations, and stakeholder representatives.[2] These local-level watershed management commissions would be responsible for the watershed's plans and targets, as well as overseeing trials of stakeholder participation and economic incentives to encourage protection of the watershed.

This on-the-ground pilot work to reform RBCs from the bottom-up could be strengthened by an exchange in which members of the large RBC that oversees the project area could work in a U.S. and Japanese river basin commission for several months. Visiting some RBC with mechanisms for involving all basin stakeholders could not only offer Chinese river managers insights into how to be more inclusive in their work, but also how to prevent or resolve water conflicts. China's problems in dealing with domestic and international water conflicts have increased in great part because government bureaucracies focus almost exclusively on managing rivers for economic development rather than following a development path to balance both human and ecological needs. In contrast, Japan, U.S. and other developed countries now put increasingly greater emphasis on ecological value of river flow. River basins are not without strife in the United States and Japan, but both countries have worked to create laws and institutions that emphasize the value of river flow and provide formal channels for participation and dispute resolution.[3]

Although many river basins in Japan and the United States are considerably smaller than the seven main basins in China, we believe they still offer lessons that could be applied at the sub-basin level in China. It merits mention since China's RBCs have been in existence since the 1950s, the reforms they are undertaking also could provide important insights for the United States and Japan, both domestically and in international assistance. The Delaware River Basin Commission (DRBC) is one U.S. RBC that merits study. Since being created in 1961, DRBC and its members (the four riparian states and federal government) have not only resolved contentious conflicts among the states, but also acted as a forum to effectively mobilize government, citizen and NGO communities to solve water shortage and pollution problems. In contrast to China's RBCs that do not have provinces as members and lack sufficient power and inclusiveness, DRBC offers an interesting model for how a commission can achieve better governance of a river basin if given sufficient regulatory authority and ability to bring together multiple stakeholders.[4]

Japanese river basin committees also offer valuable examples for Chinese counterparts, for both countries have very centralized systems of managing rivers. In 1997, Japan's River Law was amended to require the creation of river basin committees, which have since been established on many lakes and rivers. Although these committees are a relatively new

In southern China the ecosystems of many rivers and lakes are seriously impacted by eutrophication caused by pollution and agricultural runoff. This destruction of the water ecosystems often has enabled invasive species to run rampant. The above photo shows a lake in Fujian Province overrun by an invasive plant. (Photo Credit Deng Jia).

institution in Japan, they already have gained considerable experience in gathering stake-holders together and building consensus on sensitive development and environment issues in river basins. For example, the Yodo River Basin Committee is a unique consulting orga-nization in that during four and a half years it held over 400 basin management planning meetings that were open to the public.[5] Although this committee was set up by an initia-tive of the regional development bureau under the Ministry of Land and Transportation, it is run not by the bureau, but by consultation with members, who are scholars, and community and NPO representatives. A private company carries out the committee's ad-ministrative work. Opening up the discussions to the public have made them slow, for the committee is still working on draft management plan for the basin. However, once imple-mented the plan should face little opposition and conflicts will be easier to resolve.

We concur with another recommendation in the CCICED IRBM Task Force report, which suggests in addition to these on-the-ground trials more international assistance could focus on some national-level institutional and legal changes in water governance institutions in China.[6] For example, in partnership with the central government, in-ternational partners could help in reviewing and revising river basin management and water pollution control legislation to reduce contradictions and clarify institutional re-sponsibilities of RBCs. One mechanism of cooperation for revising the laws could be inter-parliamentary exchanges. For example, some members of the National People's Congress Committee on Environmental Protection and Natural Resources could meet with their Japanese and U.S. counterparts to learn about some of the more effective laws both countries have made in relation to river and water protection (e.g., the Wild and Scenic Rivers Act in the United States).

Another possible high-level initiative promoted by the IRBM Task Force to mitigate the many-headed dragon dilemma is the creation of a national-level IRBM commission that would include the National Development and Reform Commission, Ministry of Water Resources and the State Environmental Protection Administration (SEPA).[7] This commission could oversee changes in laws and create new laws to promote adoption of IRBM nationally.

Similar to many other countries, in China a crisis—such as toxic water pollution incidents in the Songhua River and major flow cut offs in the Yellow River—can act as a driving force for unifying government agencies to protect rivers. However, cooperation in response to river crises in China often has created more government centralization of authority over rivers or ineffectual campaigns—not ideal ingredients for creating an integrated institution for sustainable water governance. Creating an incentive mechanism to promote governmental collaboration would be crucial to establish strong IRBM in China. To promote such collaboration two important areas of study would be to: (1) identify specifically how and when collaboration between functional specialization organizations (e.g., water suppliers, wastewater treatment facilities, and flood control agencies) will be socially beneficial, and (2) examine how best to mobilize sufficient political will to force functional specialization organizations to collaborate when they should.[8] Information and data sharing would be the most realistic first step for China to take not only to improve planning and implementation of water protection policies, but also to save cost and time in resolving water problems, which ultimately could benefit all stakeholders.

Utilizing New Financial Mechanisms and Incentives

"Who gains the benefit and who pays the cost" are major questions when discussing the economics of sustainable river basin governance. To answer these questions Chinese policymakers appear enthusiastic about introducing market-based instruments as a new enforcement tool for environmental regulations or promoting conservation—particularly of water. Water right trading is attracting much attention among Chinese technocrats and scholars as a method of improving conservation of water. However, China's unclear water rights system and weak legal institutions hinder any current systematic application of water markets.[9]

Water rights represent a very complex and sensitive issue in China. However, both the U.S. Department of Agriculture (USDA) and the Japanese International Cooperation Agency (JICA) are independently undertaking projects focused on clarifying water rights in China. Even if a formal partnership is not feasible, the insights both the USDA and JICA gain from pilot projects and research in this area should be shared and disseminated more widely. Such information sharing could help identify some opportunities for joint work in this area.

Water pricing needs to increase in China in order to cover the costs of building and operating water supply systems and wastewater treatment plants. Currently there does not appear to be a good model of cost sharing for water conservation in China, which represents an urgent and challenging task for China's sustainable river basin governance. The government and NGO sectors in Japan and the United States could work with the

Chinese government to create funding mechanisms to promote conservation of river and wetland resources. Although the interest within the Chinese government is great, there are few Chinese or international initiatives focused on financing to promote conservation of water resources (e.g., payment for environmental services schemes, green taxes, revolving funds, and municipal bonds) or using market mechanisms (e.g., water trades and water banks). In light of these challenges in water financing, possible areas of U.S.-Japan-China collaboration are presented below.

Establishing payment for environmental services (PES) pilots. How to set up mechanisms to motivate downstream water users to compensate those upstream to protect watersheds is a challenge faced in many countries. The United States can provide examples on how this has been done successfully and one model can be found in USAID programs that have supported numerous PES pilot projects in rivers in developing countries. Another model can be found in new taxation schemes aimed at promoting protection of upstream water resources and forests that were adopted by many Japanese prefectural governments. This "green" taxation scheme only has been in use for two years in Japan, so it is too early to evaluate its effectiveness on protecting the rivers. However, Chinese policymakers and river managers could study these schemes to learn how to introduce economic incentive methods to protect water resources based on partnerships among upstream and downstream stakeholders. There is some precedent for such schemes in China in that the Chinese government has developed some compensation policies (mainly fees, subsidies, taxes, and punitive measures) to protect forestry resources. For example, timber cutting is banned in much of southwest China and reforestation is being encouraged by compensating farmers with extra grain and subsidies for converting agricultural land on slopes into forests. These forest protection programs, however, are government, rather than market-led programs.

Expanding water user associations for both water quantity and quality protection. While the World Bank has helped establish nearly 2,000 water user associations in China, many local water bureaus in rural China have faced severe difficulties in providing adequate service and in assessing sufficient water charges. Thus there is clearly need for an examination of other successful models of water user organizations overseas, particularly those that promote pollution control. For example, in the Netherlands, water board organizations that are constituted of local stakeholders have played a crucial role of setting water pollution charge rates to share the costs among members.[10]

Creating revolving funds to support water conservation and pollution control initiatives. In 1987 when the U.S. Congress amended the Clean Water Act, an innovative Clean Water State Revolving Fund (CWSRF) program was created. The CWSRF program is available to fund a wide variety of water quality projects including non-point source pollution, watershed protection or restoration, and estuary management projects, as well as more traditional municipal wastewater treatment projects.[11] Such a fund has been the target of some pilot projects in China. For example, in 2004, as part of their wetlands restoration project in the mid-reaches of the Yangtze River, WWF-China set up a revolving fund that aims to help farmers in Qiuhu village who lost land to the restoration of wetlands develop alternative livelihoods. The seed funding has helped these farmers create bamboo nurseries, sustainable fishing, ecotourism ventures, and hydroponic vegetables. In the first round of funding 104 households have paid off their loans with interest into a pool that will enable other farmers to take out loans.[12]

Utilizing municipal bonds to fund wastewater treatment plants. In terms of water pollution control, the central government does not have any formal policy to cover part of the costs for wastewater treatment plants and local governments often are unwilling to make the investment into wastewater treatment for it is viewed as a hindrance to local economic development. One possible solution that would demand considerable changes in the legal and financing institutions in China would be the creation of municipal bonds to fund environmental infrastructure such as wastewater treatment plants. The World Bank and the U.S. Trade Development Agency completed a successful pilot project in Shanghai on issuing a municipal bond for wastewater treatment. Other similar pilots could be done in other cities in collaboration with municipal governments and the National Development and Reform Commission in order to help identify the kinds of adjustments needed in existing tax and finance laws to encourage investors to enter the municipal bond market and to create mechanisms and laws to minimize risks of municipal bonds.[13]

Experimenting at small-scale water trades. While some, albeit illegal, water trades have taken place in China, as mentioned in Part II, the U.S. and Japan could build on their current work on water rights to help set up some institutions at the local level for water trading.

Incorporating broader costs and benefits into EIAs and planning. International initiatives focusing on various financing initiatives could follow the CCICED IRBM Task Force recommendation by undertaking pilot projects in a single river basin.[14] Ideally, RBC development and planning decisions should be based on environmental impact assessments (EIAs) that include criteria that value not just economic costs, but also social and environmental costs and benefits. Joint U.S.-Japanese-Chinese teams could undertake studies and pilot projects to identify obstacles and potential solutions for incorporating true environmental, ecological and social impact assessments into river management.

Opening up the Floor to Greater Stakeholder Participation

Since the mid-1990s the Chinese government has actively encouraged more public participation in the environmental sphere—not just due to international projects that promote collaboration between government and citizens, but also because of political changes during the reform period granting more openness in society. Some reforms that have spurred more citizen involvement in environmental protection and markedly changed the relationship between citizens and the state include: a more open news media, registration rules that permit individuals to establish NGOs, the right to sue in cases of personal injury, public comment requirements on EIAs, and a gradual increase in access to information. Such trends are promising and open up opportunities for international involvement in increasing public participation in river basin management. Parts II and III included some examples of domestic and international NGO and bilateral aid work in promoting public participation, but SEPA and other Chinese agencies acknowledge more such work is needed in this area. The basic requirement for public participation in river management is the ability of river basin stakeholders to access all information and have a voice in shaping IRBM planning, EIAs, and management decisions made in the basin. There are a number of opportunities for U.S. and Japanese governments, NGOs, and research centers to help push forward greater stakeholder participation in both water management and pollution control spheres.

(1) Basin-level forum. The CCICED IRBM Task Force recommended the creation of a Development and Conservation Forum in each large river basin to act as a platform for communication and consensus building between different provinces and between government, NGO, and research stakeholders.

(2) Sub-basin or municipal-based forums to promote corporate social responsibility (CSR). While industries are often the main polluters, it is difficult to motivate them or local governments to work with communities, NGOs or universities on adopting pollution prevention measures. However, some international organizations could create initiatives to educate businesses and other stakeholders in a sub-basin or city on how CSR work protecting water resources ultimately is profitable for businesses. Examples of CSR for water pollution control include: voluntarily exceeding government emission standards; disclosing emissions information to the public; mandating green supply chains; building NGO-business partnerships; participating voluntarily in water pollution trading pilot projects; and adopting transparent emergency management system mechanisms.[15]

(3) Basin-to-basin exchanges. Basin management officials and NGOs could participate in exchanges to visit Japan and the United States to learn about how the public is brought into the management of rivers and watersheds.

(4) Public hearings for basin management decisions. In areas with IRBM pilot projects one crucial mechanism will be the creation of regular participation for community members in the planning and implementation of watershed management measures. Many hearings today in China are simply meetings for the public to make comments after most discussion has taken place. In November 2005, SEPA requested international advice on designing regulations to advance public participation in EIAs. Soon SEPA will need assistance in conducting trainings once they set these new regulations.

(5) Legal assistance for pollution victims. Helping citizens access the courts in cases of water pollution accidents could play a major role in putting pressure on local governments and industries to enforce existing water pollution protection laws. Currently only one Chinese NGO is doing work to help victims of pollution, which means this legal avenue is very underserved in China.

(6) Training of Chinese NGOs. Chinese environmental NGOs have built their capacity and increased their effectiveness in great part due to support from international

NGOs, foundations, and other governments.[16] This external assistance has been invaluable, yet more could be done specifically on water. For example, there are a number of bilateral and multilateral river basin initiatives but few have brought NGOs into the process. Involving NGOs in such projects would be crucial for establishing them as legitimate participants in river basin management.

(7) Cultivating stewardship. While promoting partnerships between the government and public on jointly managing rivers is crucial, a more significant change would be to empower citizens to become stewards of the water and surrounding land. The more citizens are involved in caring for river basin resources the less costly it will be for the government to protect rivers. There are examples of the Ramsar Center Japan (RCJ) working in estuaries in Japan and India, in which RCJ not only facilitated successful citizen partnerships with government agencies, but also empowered citizens to become the leading decision-makers on how to restore the coastal lagoon ecosystem and simultaneously improve their livelihoods.[17] A stewardship approach to water resource management is a wise complement to the regulatory approach, and represents the best hope for achieving the long-term vision and sustained action needed to maintain essential ecosystem goods and services at the basin-scale over time.[18]

Conclusion

As a strong emerging economy and the "world's factory," China's impact on the global market—both in product exports and commodity imports—is great and will grow over the coming decades. The choices the Chinese government makes today in terms of environmental protection and energy conservation also will have global impacts well into the future. China has shown considerable progressiveness in its environmental and energy conservation lawmaking that could turn the country into a model for sustainable development. However, implementation and enforcement of these laws have been very uneven, particularly in the water sphere.

While the challenges facing China in the water sector are formidable, the openness of the Chinese government for internal reforms on water management and interest in outside models to mitigate water problems highlights an important opportunity for Japan, the United States, and other countries to assist China. The United States and Japan could offer China different backgrounds and experiences in river basin management, which could inspire many options and ideas for protecting Chinese rivers. Such collaboration also could help encourage new U.S.-Japan water partnerships in other countries. In short, we believe Japanese-U.S. collaboration in reaching across the Pacific to promote sustainable river basin governance could not only contribute to water security in China, but also around the world.

Notes

1. USAID Environment Web pages: http://www.usaid.gov/our_work/environment/water/index.html
2. CCICED. (2004). *Promoting Integrated River Basin Management and Restoring China's Living*

Rivers (CICED Task Force Report on Integrated River Basin Management). Beijing: CCICED.

3. Kataoka, Naoki. (2005). "Conservation of the Waterfront Environment among Japan's Rivers: Institutions and Their Reforms of River Basin Management." In Jennifer L. Turner and Kenji Otsuka (Eds.), *Promoting Sustainable River Basin Governance: Crafting Japan-U.S. Water Partnerships in China. IDE Spot Survey No. 28* (pp. 37- 46). Chiba, Japan: Institute of Developing Economies/IDE-Jetro. [Online]. Available: http://www.ide.go.jp/English/Publish/Spot/28.html. Collier, Carol R. (2005). "Sustainable Water Resources Management in the United States: Use of River Basin Commissions to Promote Economic Development, While Protecting the Environment and Improving Community Quality." In Turner and Otsuka (Eds.), op cit. (pp. 47-59). Nakayama Mikiyasu. (2005). "China as a Basin Country of International Rivers." In Turner and Otsuka. (Eds.), op cit. (pp. 63-71).

4. Collier, Carol. (2005).

5. Nakamura, Masahisa. (2005). "Experiences and Problems of River Basin Management in Biwa Lake-Yodo River." *Ajiken World Trend,* No.122, 26-30. [In Japanese].

6. CCICED. (2004).

7. CCICED. (2004).

8. Wolff, Gary. (2005). "Economies of Scale and Scope in River Basin Management." In Jennifer L. Turner and Kenji Otsuka (Eds.), op cit. (pp. 73-82).

9. Wang Yahua. (2005)."River Governance Structure in China: A Study of Water Quantity/ Quality Management Regimes." In Jennifer L. Turner and Kenji Otsuka (Eds.), op cit. (pp.23-36).

10. Fujita, Kaori. (2005). "Evaluating Cost Sharing for Sustainable River Basin Management: Case Studies in Netherlands and Japan." In Jennifer L. Turner and Kenji Otsuka (Eds.), op cit. (pp. 103-122).

11. U.S. EPA's Clean Water State Revolving Fund Homepage. http://www.epa.gov/owmitnet/cw-finance/cwsrf/basics.htm

12. "Revolving Fund Encourages Wetland Wise Use Initiatives in Qiuhu Village." (2005). *WWF HBSC Yangtze Programme Newsletter.* Vol. 2 April 1-June 30. [Online]. Available: http://www.wwf-china.org/english/loca.php?loca=91

13. Baldinger, Pamela and Jennifer L. Turner. (2004). *Municipal Financing for Environmental Infrastructure in China* (Final Report).

Washington, DC: Woodrow Wilson Center's China Environment Forum. [Online]. Available: www.wilsoncenter.org/cef

14. CCICED. (2004).

15. Hua Wang, Jun Bi, David Wheeler, Jinnan Wang, Dong Cao, Genfa Lu, and Yuan Wang. (2003). *Environmental Performance Rating and Disclosure: China's GreenWatch Program* (World Bank Working Paper # 2889). [Online]. Available: http://www.worldbank.org/nipr/china/Greenwatch-JEM-1.htm. IGES. (2005). *Information Access as a Vehicle for Sustainable Development in Asia— Establishing a Regional Agreement in Asia* (Policy Brief #002). [Online]. Available: http://www.iges. or.jp/en/pub/pb002.html

16. Turner, Jennifer and Lü Zhi. (2006). "Building a Green Civil Society in China." In Linda Starke (Ed.), *State of the World 2006.* (pp.152-170). Washington, DC: Worldwatch Institute.

17. Nakamura, Reiko. (2005). "Essentials of Stakeholder Participation in the Wise Use of Wetlands: Good Practices of Two Lagoons in Japan and India." In Jennifer L. Turner and Kenji Otsuka (Eds.), op cit. (pp. 141-151).

18. Volk, Richard. (2005). "Fostering a Stewardship Approach to Water Resource Management." In Jennifer L. Turner and Kenji Otsuka (Eds.), op cit. (pp.153-167).

Woodrow Wilson
International
Center
for Scholars

水パートナーシップの構築に向けて

－中国における持続可能な流域ガバナンスを促進するための国際協力－

同舟共済

可持续流域治理的国际合作

ジェニファー・**L**・ターナー／大塚健司

吴岚・大冢健司

2006

Available from the China Environment Forum
Woodrow Wilson International Center for Scholars
One Woodrow Wilson Plaza
1300 Pennsylvania Avenue, NW
Washington, DC 20004-3027

www.wilsoncenter.org/cef

リサーチ・アシスタント：リンデン・エリス、ティモシー・ヒルデブラント、
シャルロッテ・マックオースランド、ルイーズ・ユエン、ルル・ヂァン
中国語訳：セレーナ・イイン・リン
日本語訳：大塚健司
制作編集：リアン・ヘルパー

研究助理： Linden Ellis, Timothy Hildebrandt, Charlotte MacAusland, Louise
Yeung, and Lulu Zhang
日文译者： 大冢健司
中文译者： 林依莹
编辑者： Lianne Hepler

ISBN 1-933549-06-8

巨大ダムの建設が計画されている虎跳峡の渓谷（写真：馬軍）。
虎跳峡上的山崖。此处一个大型水坝正在筹划之中。

ウッドロー・ウィルソン国際学術センター

ウッドロー・ウィルソン国際学術センター（ウッドロー・ウィルソンセンター）は、優れた研究を支援し、その研究をワシントンの政府関係者の関心につなげ、アイデアの世界を政策の世界に結びつけることを目的としている。センターはウィルソン大統領への政府および国民による記念として1969年に議会により設立された。センターは国際フェロープログラムを助成し、独立した評議会（大統領から任命された10名の市民と国務長官を含む9名の政府官僚から構成される）を有し、無党派で、いかなる公共政策に対しても特定の立場をとらない。センターの会長はリー・H・ハミルトン閣下、理事長はジョセフ・B・ギルデンホーンである。www.wilsoncenter.org

中国環境フォーラム

1997年以来、ウッドロー・ウィルソンセンター中国環境フォーラムは、情報の共有、政策対話の促進、そして最も重要なのが、アメリカ、中国および他のアジア諸国の政策立案者、NGO、研究者、実業家、ジャーナリストとの間で共通の環境・エネルギー問題を解決すべくネットワークを構築することにより、中国における持続可能な発展への道のりを模索してきた。中国環境フォーラムは常に多様なバックグラウンドと所属機関−主な合衆国政府機関、エネルギー、中国研究、外交政策、経済と貿易、環境、農村開発の各分野−からの専門家を呼び集めている。中国環境フォーラムは、月1回の会合を通して、中国における環境と持続可能な発展に関する最も重要な問題をテーマとして定め、創造的なアイデアを紡ぎ出し、また官民協力の機会を提供することを目的としている。www.wilsoncenter.org/cef

アジア経済研究所

アジア経済研究所は、1960年（昭和35年）通商産業省（現経済産業省）所管の特殊法人として、 開発途上国・地域の経済、政治、社会の諸問題について、基礎的・総合的研究を行う社会科学系研究所として設立された。1998年7月、日本貿易振興会（ジェトロ）と統合、 2003年10月からは独立行政法人日本貿易振興機構アジア経済研究所としてアジア、中東、アフリカ、ラテンアメリカ、オセアニア、東欧諸国などすべての開発途上国・地域との貿易の拡大及び経済協力の促進を図ることを目的に調査研究、成果普及、国際交流事業を実施している。www.ide.go.jp

著者紹介

ジェニファー・L・ターナー：1999年よりウッドロー・ウィルソン国際学術センター中国環境フォーラムのコーディネーター。中国の環境問題解決に向けて、アメリカ、中国、そして他のアジア諸国の政府、NGO、研究、実業界の間における対話と協力を促進することを目的として、会議やスタディツアーを統括し、出版活動を行っている。1997年にインディアナ大学にて比較政治学と環境政策のPhDを取得。現在、中国における水関連政策と"グリーンな"市民社会の発展について研究を行っている。

大塚健司：日本貿易振興機構アジア経済研究所新領域研究センター環境・資源研究グループ研究員。1993年より同研究所にて中国の研究者らと中国の環境問題に関する共同研究を行っている。主に中国における環境政策過程、水問題、環境紛争、環境意識、コミュニティ、NGOに関する現地調査を踏まえた研究を実施。1992年に筑波大学大学院環境科学研究科にて修士号（環境科学）を取得。

伍德罗·威尔逊国际学者中心

伍德罗·威尔逊国际学者中心通过支持卓越的学术研究并将研究成果在华盛顿参与实事政治的政府官员间传播，来达到联合学术界和政策界的目的。美国国会在1986年建立了这个无党派的学术中心，作为国家对威尔逊总统的正式纪念。该中心设有13个项目，支持国际学者项目，并且具有独立的理事会（由总统指派的10位公民，包括国务卿的9位政府官员组成）。中心的主席为尊敬的Lee H. Hamilton。Joseph. B Gildenhorn为理事会的主席。www.wilsoncenter.org

中国环境论坛

从1997年开始，中国环境论坛通过促进信息共享、鼓励政策讨论，最重要的是通过建立中美以及其它亚洲政策决策者，非政府组织，学者，工商业以及新闻人士间的联系网络，寻求在中国建立可持续发展的方法，以解决广泛的环境和能源问题。中国环境论坛定期邀请不同背景及组织的专家相聚，包括能源，环境，中国研究、经济学既农村发展领域的美国和国际专家。中国环境论坛通过每月举行的会议和每年一期《中国环境期刊》，来辨认中国最重要的环境和可持续发展话题，并探寻有创造性地想法和政府及非政府部门的合作机会。www.wilsoncenter.org/cef

日本贸易振兴机构亚洲经济研究所

亚洲经济研究所成立于1960年，是通商产业省(现经济产业省)所法定成立的组织，主要从事发展中国家与区域在经济，政治与社会议题的全面基础社会科学研究。从1998年七月与日本贸易振兴会合并以后，亚洲经济研究所的目标是促进日本与其它发展中国家和区域之间的经济合作与贸易发展，包括亚洲，中东，非洲，拉丁美洲，大洋洲和东欧。自2003年10月1日起，日本贸易振兴会成为一个独立的行政法人机构，日本贸易振兴机构。www.ide.go.jp

关于作者

从1999年以来，吴岚(Jennifer Turner) 担任威尔逊国际学者中心中国环境论坛的协调人。作为协调人，她举办会议，组织 研究考察团，发行出版物，旨在促进美国，中国和其它亚洲国家的政府，非政府组织，研究机构和企业部门之间的对话与合作，以协助解决中国的环境问题。1997年她取得印第安纳大学比较政治与环境政策的博士学位。近年来她的研究着重于水资源政策以及在中国日渐成长的 "绿色" 公民社会。

大冢健司 (Kenji Otsuka)是日本贸易振兴机构亚洲经济研究所跨领域中心环境与资源研究组的研究员。从1993年开始，他开始与中国的专家学者进行研究项目，相关研究包括环境政策的实施过程，水资源与流域管理，环境污染纠纷，环境意识宣导，和中国社区/非政府组织的环境运动。1992年于筑波大学取得环境科学的硕士学位。

目次

目录

まえがきと謝辞

　中国が直面する水問題に関する本リポートは、ウッドロー・ウィルソン国際学術センター中国環境フォーラム（CEF、ワシントンDC）のジェニファー・ターナーと日本貿易振興機構アジア経済研究所（IDE—JETRO、千葉）の大塚健司による共同プロジェクトをもとに作成されたものである。その共同プロジェクト「日米水パートナーシップの構築に向けて：中国における持続可能な流域ガバナンスの促進のために」（アジア経済研究所における事業名は「中国の持続可能な流域管理と国際協力－日米水協力イニシアティブによる展望」）は国際交流基金ニューヨーク日米センターの助成を受けた。このプロジェクトにおいて、3カ国の研究チームメンバーは各リサーチペーパーを執筆し、その論文集は*Promoting Sustainable River Basin Governance: Crafting Japan－U.S. Water Partnerships in China,* IDE Spot Survey No.28として2005年3月に出版された。また、日本の研究チームメンバーを中心に日本語のレポートを執筆し、その成果は2005年11月にアジア経済研究所の月刊情報分析誌『アジ研ワールドトレンド』特集号「中国における持続可能な流域ガバナンスと国際協力」として刊行された。このリポートは主としてIDE Spot Surveyのいくつかの研究成果を取り込んでいるが、大部分が新たな情報を含んだものとなっている。

　このリポートの作成にあたっては多くの個人と団体の助力を得ている。何よりもまず中国、日本、アメリカにおける3回にわたる流域ガバナンスのスタディツアーに積極的に参加していただいた10人のメンバーに感謝を申し上げたい。そのメンバーは、キャロル・コリアー、ゲイリー・ウルフ、藤田香、胡勘平、片岡直樹、中村玲子、リチャード・フォーク、王亜華、山田七絵、于暁剛の各氏である。メンバーは、スタディツアーを通して、中国が流域管理における3つの重要な領域－流域管理機構・制度、資金調達、公衆参加－における日米の経験を踏まえ、いかに統合的流域管理を実現するかという問いにこたえるべく、各々がリサーチペーパーの執筆を行った。もう一人の重要なメンバーは、博士課程への進学のためウィルソンセンターを離れた中国環境フォーラムの元アシスタント、ティモシー・ヒルデブラント氏であり、彼はスタディツアーを構想および組織する上で重要な役割を果たし、このプロジェクトの成功にとって欠かせない存在であった。日本の研究会におけるもう一人のメンバーである中山幹康氏には中国の国際河川に関する貴重な論文を執筆いただいた。また中村正久氏と北野尚宏氏には東京における国際ワークショップにおいて報告・コメントをいただいた。あわせて感謝申し上げたい。

　スタディツアーにおいて数え切れない人びとが私たちのグループとともに席に着き、いかに効果的に流域管理を行うかについて、それぞれの経験と見識の共有を図った。以下の団体の個人から私たちのリサーチチームにいただいた助言に謝意を表したい（アルファベット順）。アサザ基金、チェサピーク・ベイ基金、チェサピーク・ベイ・プログラム、中国環境・発展国際協力委員会、中国環境・持続可能な発展文献研究センター、中国水利部、コンサーベイション・インターナショナル北京事務所、デラウェア流域委員会、中国スウェーデン大使館、中国スイス大使館、EU北京事務所、中国緑色時報、中国緑色報、グリーン・リバー、グリーン・ウォーターシェッド、GTZ、海河水利委員会、国際建設技術協会、ア

ジア経済研究所、ポトマック河州際委員会、国際協力銀行、国際協力機構、水資源機構、神奈川県総務部税務課税制企画担当、メリーランド州自然資源部、桃山学院大学、ニューヨーク市地域計画協会、中国人民大学、太平洋研究所、ラムサールセンター(RCJ)、天津市環境保護局、東京経済大学、東京都港湾局防災事務所、清華大学、筑波大学、英国国際開発省（DFID）、東京大学、米国国際開発庁（USAID）、米国陸軍工兵隊、米国環境保護局、ウェットランド・インターナショナル、ウッドロー・ウィルソンセンター、世界銀行北京事務所、ワールド・フィッシュ・センター、WWF中国。

　私たちはまたこのリポートの作成にあたり、情報提供をいただき、またドラフトを読んで貴重な示唆を賜った、バルク・ボクサー（スタンフォード大学）、エリザベス・エコノミー（米国外交委員会）、パトリック・フレイモンド（中国スイス大使館）、ピング・ホイディング（中国スウェーデン大使館）、ブライアン・ローマー（米国農業省）、リチャード・フォーク(USAID)、ジェームス・ニッカム（東京女学館大学）、王亜華（清華大学）、馬軍（『中国水危機』の著者）、温波（パシフィック・エンバイロメント）、フェンシ・ウ（香港中文大学）、森尚樹（日中友好環境保全センタープロジェクト日本専門家）、石渡幹夫（国際協力機構）、片岡直樹（東京経済大学）、藤田香（桃山学院大学）、山田七絵（アジア経済研究所）の各氏に謝意を表したい。また、ウィルソンセンターの中国環境フォーラムにおける3名のリサーチ・アシスタントで、最新情報や新聞報道の収集のみならず、重要かつ時間のかかる編集作業をしていただいた、リンデン・エリス、シャルロッテ・マックオースランド、ルイーズ・ユエン、ル・ヂァンに謝意を表したい。最後に、このプロジェクトが実り多いものになるよう取り組む意欲を私たちに与え、スタディツアーの一部に参加し、また東京にて国際会議場を提供していただいた国際交流基金日米センターからのたゆまぬ支援、とくに、キャロリン・フィッシャー、原秀樹、茶野純一、佐藤敦子の各氏による助力に感謝申し上げたい。

　このリポートの作成にあたり、上記個人およびウッドロー・ウィルソンセンターやアジア経済研究所のスタッフから多大な貢献をいただいたが、内容についての責任はすべて私たち筆者にある。本リポートで表明された見解は筆者個人に属し、ウッドロー・ウィルソンセンターあるいはアジア経済研究所による公式見解ではない。

エグゼクティブ・サマリー

　中国は多くの水危機に直面している。規制を逃れた工場や未処理の都市下水から排出される化学物質に汚染された湖沼や河川、地下水や表流水の過剰採取による深刻な水不足、森林や湿地の破壊による洪水などである。中国における水質の悪化と不足は、人口移動、健康リスク、食糧安全保障問題を引き起こしている。水問題は結果として中国の社会、経済、政治の安定性に影響を及ぼす可能性がある。

　中国の水問題の中心的課題として河川生態系の保護が必要である。中国の河川への脅威を緩和する必要性は、法制度、政策、プロジェクトを強化し、統合的流域管理とより包括的な汚染防止戦略を促進する国内および国際的な努力を促している。これまで統合的流域管理を実施するための中心的戦略として中国政府は流域委員会制度の改革を図ってきた。このトップダウン方式は河川管理を抜本的に改革するための鍵を握っているが、同時に市民やNGOが河川における開発と保護の政策決定と監視に参加できるようエンパワーメントを行うことが重要である。いくつかの国際環境NGOが中国において河川流域保護プロジェクトを立ち上げ、政府機関、コミュニティ、中国のNGOを集め、地域の河川を保護するための多様なステークホルダーの参加によるプロジェクトを形成している。

　日米の政府とNGOは中国においてそれぞれ水や河川保護に関するプロジェクトを実施している。しかし、これらのプロジェクトの多くは小規模で期間が短いため、中国における真の統合的流域管理のために必要な制度改革を促す可能性は限定されている。中国の統合的流域管理に対しより大きな影響を及ぼすために、日米が共同で、流域管理、資金調達、ステークホルダーの参加といった分野において事業を実施することができるであろう。

　このリポートは日米（および他の国々）における政府、NGO、研究セクターが中国における流域ガバナンスに関する共同プロジェクトを実施するうえでのいくつかの選択肢を提示することを目的としている。中国における水に関するさらなる国際協力に関する議論のステップとして、まず第1章で、中国における水問題の重大性について議論する。次に第2章において現行の水関連の法制度の効果とともに、中国において水問題に取り組む小規模ではあるが成長しつつあるNGOの活動をレビューする。第3章において、中国における持続可能な水資源管理を促進するための国際協力を概観するとともに、そのギャップに焦点をあてる。第4章の結論において、日米の政府、NGO、学術界が中国における持続可能な流域ガバナンスを促進するために共同で（あるいは並行して）取り組むことができるテーマを提示する。

　以下、日本語訳は英語本文の抄訳である。全訳についてはウッドロー・ウィルソンセンター中国環境フォーラムのホームページwww.wilsoncenter.org/cefを参照されたい。

第1章　危機にある中国の河川

　過去25年にわたり、中国経済の奇跡は何百万人もの人びとを貧困から救い出したものの、環境に対する負のコストももたらした。中国の環境問題に関する統計からは厳しい状況がうかがえる。世界における最も汚染された20都市のうち16都市が中国にある。中国はすでに米国に次いで多くのエネルギー（ほとんどが低質炭）を消費し、多くの温室効果ガスを排出しており、今後20年間でいずれも米国を追い抜く見込みである。中国全土の3分の2が（同時に韓国、日本も）石炭燃焼に起因する酸性雨の影響を受けている。全国の動植物の20パーセントが危機にさらされている。北方地域では、水不足により砂漠と化した農地を放棄せざるを得ない生態移民が発生している。中国の都市を流れる河川の75%が飲用や漁業に適さない。国家環境保護総局の潘岳副局長も、中国の環境悪化の重大性に言及し、環境悪化はGDP年成長率のおよそ8%に相当する損失をもたらしており、それによって経済の奇跡はもはや神話となったとしている。多くの環境問題のなかで、深刻な水不足、拡大する水汚染、河川生態系の管理不全は中国の経済、生態系、そして人びとの健康への主な脅威となっている。

水の汚染

　中国における主な河川はすべて深刻な汚染の影響を受けており、それによって人々の健康は脅かされ、工業生産は停止に追い込まれ、また河川生態系は破壊されている。工業への弱い規制や不十分な水資源の総合管理が、中国における深刻な水汚染問題を引き起こす主たる制度的失敗の2つである。2002年以来、約630億トンの廃水が毎年中国の河川に流れ込み、そのうち62%が工業汚染源からの廃水であり、38%がほとんどあるいは全く処理されていない都市生活汚水である。中国の主な河川－とりわけ淮河、海河、黄河—沿いの地域社会では、土壌や水の高レベルの汚染により、癌、腫瘍、自然流産、知的障害の発症率が異常に高いことが報告されている。

水不足

　中国の年間1人当たり水供給量は世界平均より25%低く、2030年には1人当たり供給量が現在の2,200m³から1,700m³以下と、世界銀行が水不足国と定義する水準になると予測されている。

　水の不足と汚染がもたらす農業セクターの損失額の推定値は、世界銀行の240億ドルから中国の新聞報道による82億ドルまでの幅がある。水不足は北方地域で最も深刻であるが、全国レベルで持続可能な発展を阻害する主な要因となっている。中国における640の主要都市のうち400都市が水不足に直面しており、2003年には280億ドルの工業生産額の損失を招いたとされている。中国の農村地域では約6,000万人の人々が日常生活に必要な水を十分に得られていない。増大する水不足は黄河において顕著にあらわれており、1990年代中頃から水不足のために年間200日間も河川の水流が海に届かない断流がしばしば発生している。

水生態系の退化

　水汚染と過剰取水が流域の悪化の主な原因であるが、多くの河川－とりわけ長江－は森林破壊、湿地の干拓、洪水地域における不適切な開発やインフラプロジェクトによっても脅かされており、それらがさらに洪水の規模と被害を拡大させている。さらに不十分な計画に基づく長江の水利プロジェクトは、河川の自然流を妨げ、流域の生態系を脅かし、河川の生物多様性と生産性にかなりの損失を与えている。

　中国では水生態系の退化、汚染、不足は環境的および経済的コストをもたらすだけではなく、社会の安定にも悪影響を与えている。1990年代中頃、ある中国共産党中央の報告は、環境悪化と汚染は中国における社会不安の4つの主な要因のひとつであることを認めている。

統合的流域管理（IWRM）における3つの中心的要素

　中国の河川が直面する問題を検討することにより、中国で拡大している水危機に加え、効果的な流域ガバナンスを妨げる政治・社会的問題を理解することが可能となる。世界のほとんどの国が水資源保護における複数の課題に直面しているが、中国は水資源、とりわけ河川の持続可能な管理において重大な失敗をしている。無秩序な開発と十分な調整が行われていない水資源管理制度による不利な影響は、中国が河川管理により包括的なアプローチ－とくに統合的流域管理－を適用することの必要性を明示している。統合的流域管理に関する複雑な概念のなかで、(1)流域管理機構、(2)資金調達メカニズム、および(3)公衆参加の3つが、中国の政策立案者、NGOおよび国際援助機関が中国におけるよりよい流域ガバナンスを促進するうえでまず強調すべき重要な制度的要素であると考えられる。以下、これら3つの制度的要素について簡単に見ておこう。

ばらばらの管理体制

　調整が不十分で効果的ではない水資源管理が中国における水問題の核心にある。中国最初の水法（1988年制定）および関連する規則によって、水資源保全のための各種措置（水費の徴収、水量配分計画、取水許可、節水器具の導入など）が採られるようになった。しかし、地方政府および水利委員会によるモニタリングや執行能力の弱さに加えて、明確な水利権の確立の難しさが、これら多くの水資源管理の改革の妨げになっている。2003年に改正された水法の主な目的は、水資源の保全と管理の手段の実施を改善するために、水利委員会の権限を強化することであった。

　中国七大河川の水利委員会はもともと水資源の開発、発電、洪水被害の緩和、航路施設の整備のために1950年代に設立された。水利委員会は水利部の派出機構として、強力な技術および水文の専門家を抱えているが、水資源保護と保全に関する措置を実施し、モニタリングする管理能力に欠けている。一般に、中国の水利委員会は河川生態系の健全性を十分重視していないか、上下流の需給調整に十分取り組んでいない。また、他の政府機関、省および地方政府との不十分な（そして時には敵対する）連携のために水利委員会の有効性が限定されている。さらに水利委員会のもう一つの欠点は、委員会が流域管理の取り組みに広範なステークホルダーを取り込むことができないことである。

限定された資金調達手段

　河川の水利用と導水プロジェクトの計画において、水利委員会および他の政府機関は経済、社会、および環境面でのコストをあまり考慮していない。中央政府は高らかにキャンペーンや5カ年計画における河川水質改善の目標をかかげ

贛江源流近くにある漁船。源流の水質は比較的きれい（写真：肖高平）。

るものの、地方政府が必要とする資金がしばしば不足している。例えば、10年にわたる淮河の環境キャンペーンはほとんど河川の水質改善に寄与しておらず、また第10次5カ年計画において多くの水質改善目標があったにもかかわらず、政府による資金調達は計画で掲げられた目標に比べてなお30%不足している状況である。

　中国では、中央政府の補助金に依存する代わりに、企業や地方政府に対して河川流域の生態系を保護するための、特に下流の水利用者が上流の水質基準に基づく費用負担をするといった、経済的インセンティブを与える必要がある。中国の水利委員会および都市は、廃水処理施設の建設に投資するためのリボルビング・ファンドや債券などの資金調達メカニズムを欠いている。そのようなメカニズムの主な障害は水使用料金の料率の低さと徴収率の低さにある。中国において、環境税、水取引市場、上下流間の環境に関する費用負担制度など、河川における水資源の保全を促進するような経済的手段の導入が遅れている。

水資源政策の決定過程における透明性と公開協議の欠如

　中央・地方両レベルにおいて、政府は市民に対して、開発プロジェクトの計画や水資源管理の方案について情報提供も協議も行っていない。水資源政策の分野において一般的に、人びとが苦情や異議を申し立てたりするような参加の機会は限られている。さらに、中国の市民は、水汚染による被害について政府に苦情を申し立てる、あるいは汚染事業者（最近では政府機関の場合もある）を訴えることは可能であるが、これらの試みは必ずしも成果をあげるとは限らない。水資源政策や水利プロジェクトのモニタリングにより広範な市民やNGOが参加することによって、中国の河川流域保護の改善が期待できる。

中国における日米水パートナーシップの可能性

　急速な工業化、生活水準の向上、多くの農村住民の農業労働からの解放をもたらした中国における経済改革は、環境状況を悪化させ、中国における人々の健康および経済に直接影響をもたらしている。そのなかで、アメリカ、日本、中国が多層な環境パートナーシップを確立する必要性は中国の世界経済への統合によって経済成長と環境悪化の両面が加速されるにしたがってきわめて重要となる。中国の水問題は深刻であるが、小流域の管理、資金調達、ステークホルダーの参加といった領域において日米が協力できる多くの機会が存在している。

第2章 河川流域の保護に関する中国国内の取り組み

　深刻化する水不足や汚染の問題に効果的に取り組むことが困難であるため、トップの指導層は水資源管理や汚染問題を扱う新たな法律の制定（そして従来法の改正）や水利委員会制度の改革に迫られ、取り組みを進めてきた。水資源保護は、持続可能な水の供給が経済成長を支える上で欠かせないことから、直近の5カ年計画において優先的投資分野となっている。これらトップダウンの方法は、水資源管理関連法や制度の改革、そして水汚染防止施設の改善にとってきわめて重要である。同時に重要なのは、政府が水セクターにおいてボトムアップの方式で市民やNGOの参加を図るための政治的スペースを拡げ続けることであろう。

中国の水問題に対するトップダウンの取り組み

ハイレベルな優先課題としての位置づけ

　第10次5カ年計画（以下10・5計画）（2001－2005年）において、政府は、とくに水問題について、高い環境保護目標値を設定しているが，8,500万ドルの投資見込みからなお30％が不足している。10・5計画において、政府は145の都市下水処理場を淮河、海河、遼河、太湖、巣湖、滇池の流域で増設することを目指している。さらに、都市下水処理目標達成率は50％となっている。中国政府は工業および都市の排水処理率を向上させることに関して進展が見られるものの、10・5計画中に建設された下水処理施設に関する国家環境保護総局の検査によると、半数しか稼働しておらず、残り半数は地方政府が運転コストが高いとして施設の運転を停止させていた。

　中国共産党は2005年10月初めに第11次5カ年長期計画（以下11・5計画）（2006—2010年）の提案を行い、その最終案は2006年3月の全国人民代表大会において承認される予定である。その計画の一部では、湿地の保護と沿海地域の破壊された生態系の回復を行う環境政策が必要であることがうたわれている。また計画には、3つの主な河川（淮河、海河、遼河）と3つの主な湖（太湖、巣湖、滇池）に加えて、三峡ダム地域と長江・黄河上流の汚染防止処理を重点とすること、南水北調の導水ルートにおける汚染防止処理を優先課題とすることなどが含まれている。さらに新5カ年計画では、国家環境保護総局はすべての都市において都市汚水処理率を現在の45％から60％に引き上げることを目指している。前国家環境保護総局長の解振華は11・5長期計画において環境保護投資を1兆3,000億元（1,566億ドル）と、GDPの1.5％以上にすることを目指していると指摘している。

中国における水量管理

　中国はかなり包括的で大部の水資源保護に関する法令を制定しているが、主要な制度的問題によって水資源、とりわけ河川の効果的な管理が妨げられている。水量管理に関するひとつの重要な課題は、水は国家に属し、オープンアクセスの資源として扱われているために、明確な水利権システムを欠いていることである。不明確な水利権は、節水を阻み、セクター間の紛争なしに（特に農業から工業への水の移転）水の移転を可能とする水取引の妨げとなっている。

地方政府は水の利用を制限することで地方経済に損害を与えることをおそれ、長い間、水資源費と取水許可制度に関する改革を妨げてきた。しかし、近年、いくつかの主要都市は水資源費を引き上げ、さらに水道メーターの導入を行い始めた。こうした試みは、地表水および地下水源からの危険な過剰汲み上げを減速させるうえで必要である。にもかかわらず、水不足の都市では、費用、許可制、あるいはその他の節水政策を採るよるも、新たな水資源を開発することに力を入れがちである。北京市はこのサプライサイド・マネージメントを推進し、首都の増大する渇きを癒している。

地方レベルにおける水資源管理を改善するための好ましい動向としては、多くの都市において水務局が設立されていることが挙げられる。この新たな部局は地方の環境、水利、建設局を統合して水資源を共同管理するものであり、いわゆる「多龍頭」問題を解決しようとする地方の制度革新といえる。別の希望的であるが、まだ法的には認められていない地方レベルでの実験として、県と県、あるいは工業と農業の間における水の取引の例がある。最初の水の取引は浙江省で2000年に行われており、義烏市（県）が上流の東陽市（県）の貯水湖から毎年5,000万立法メートルの取水権を購入したというものである。

水質管理の課題

水汚染防止処理法は1984年に制定され、1996年に改正された。この法律は各級政府の環境保護局が水汚染防止処理の監督管理を行うことを定めている。汚染を規制する法的な権限を有しているにもかかわらず、企業への地方保護主義によって環境保護局は水汚染防止処理法を効果的に運用して水汚染を防ぐことができないでいる。汚染課徴金は低く、また地方政府によりその80%が汚染企業へ還元されている。地方官僚はしばしば地方政府に多額の税金を納めている工場からの汚染物質の違法排出に目をつぶっている。

統合的流域管理の障害

2002年の水法の改正によって文面上は水利委員会の権限が強化されたものの、統合的流域管理を十分に実施できるよう、さらなる改革と能力強化が必要である。中国の水利委員会は水利部の派出機構にすぎず、流域管理に対してトップダウンの限られたアプローチしかとることができていない。水利委員会は、水質問題に関する完全な権限を欠いているだけでなく、地方政府あるいは市民による関与を可能にする制度を備えていない。事実、流域委員会というのは間違いであり、水利委員会には「委員」はおらず、流域管理のトップダウンの構造のなかで、省あるいはそれより下級の地方政府が政策形成を行ったり水量分配を行ったりするような公式のメカニズムもないのである。

環境影響評価法

数年来、国家環境保護総局は、自然資源と人々の健康を守るための新たな手法として公衆参加の拡大を提唱している。環境政策の形成および実施に人々が関与する権利は中国で最初に試行された環境保護法においてあいまいにしか認められていなかった。しかし、環境法やインフラプロジェクトに関与する市民の権利は、2003年に制定された環境影響評価（環境アセスメント）法など、最近の法制度の制定によってようやく明確化、強化されるようになってきた。1990年代に制定された環境アセスメントに関する行政法規では建設プロジェクトにしか適用されなかったが、この新たな法律ではインフラおよび他の建設プロジェクトの計画評価も求めている。とくに、環境影響評価報告書を発行し、パブリック・コメントを可能にしなければならないとされた。

国家環境保護総局と地方環境保護局は説明会や公聴会を開く明確な手続を欠いているために、アメリカ法律協会のような国際環境NGOは、公聴会や説明会、その他の形式による公衆参加メカニズムの運用に関する研修を行っている。さらに、日本国際協力機構（JICA）は国家環境保護総局が環境アセスメントにおける住民参加のガイドラインの実施細則を策定することを支援している。中国政府は国際団体が環境保護の規制およびプロジェクトを改善するための支援を歓迎しているだけでなく、国内のNGOによる環境保全活動や環境教育に対してますますオープンになってきている。

中国の水問題に対するボトムアップの取り組み

表舞台に立つ環境 NGO

　1994年以来、中国の環境政策決定における人々の役割は、社会団体（NGOなど）の登記管理制度のもと、増大してきた。中央政府指導層が市民社会への政治的スペースの拡大を認めたのは、急速な経済成長と崩壊しつつある（計画経済時代の）社会福祉システムにより増大する社会問題や環境問題に対応する上で、政府は市民からの広範囲な支援を必要としたからである。しかし、中国のNGOは政府主管部門の後ろ盾を必要とし、支部を作ってはいけないなど登録への規制はかなり厳しく、この政治的スペースの拡大はなお限定的である。他に、NGOに対する規制として、同じ市や省に同じ分野の団体は1つに限られるという規定もある。

　1994年にこの新たな規制のもとで登記した最初の草の根環境NGOは自然の友である。他の環境団体は登記しようにもできず、企業として登記するか、正規の地位がないまま運営されていることがしばしばである。登記管理制度を避けるために、インターネットグループとして設立される環境団体が増加している。現在、環境NGOは中国で2,000団体近くあり、市民社会の発展の先駆者となっている。当初、中国の環境NGOは学校における環境教育の推進やリサイクル、節水、野生動物の保護などの問題を一般の人々に知らせるなど、「安全な」活動を行う傾向があった。非常に数が少ないものの、中国のNGOのなかには、流域や河川の保護に関する事業に取り組む団体があり、その事業の多くには公衆参加が含まれている

　2004年から2005年にかけて、中国の環境NGOの発展は転換点を迎えた。中国で自然河川が残っている2つの河のひとつである怒江の雲南省域内において、13基の水力発電ダムの建設計画が明らかになり、環境運動家とジャーナリストがプロジェクトの情報公開を求めて全国的なキャンペーンを行ったのである。この大規模な論争は温家宝首相の注意をひき、2005年2月に環境アセスメントを棚上げしてダム計画を一時中止させた。2005年8月、広範囲の国内のグループ（NGO61団体と99人の研究者および政府官僚）は、この自然水流のある河、怒江における水力発電計画について、政府がダム建設を許可する前に、環境アセスメントの情報公開を急ぐべきであるとする公開書簡を中央指導層に提出した。

　怒江をめぐるこのNGOの活動は単なる「ダム反対」運動ではなく、むしろ中国における水資源管理と環境政策形成に透明性と市民参加の拡大を求める運動である。

水問題への取り組みにおけるトップダウンとボトムアップの統合

　あらゆる国々が水資源を保護するための法令の実施に苦しんでいるが、とくに中国の直面する障害は、人口圧力、急速な経済成長、官僚のなわばり争い、不明瞭な水利権、地方保護主義など、困難なものばかりである。中国政府は水

工場からの硫黄および他の汚染物質の排出によって—長江支流の大渡河にあるこれらの工場のように—長江における浸食、汚水およびごみなどの水質問題がさらに悪化している（写真：楊欣）。

汚染を防止するための強力な法規制を打ち立て、水資源を保全するための法令を強化してきた。さらに、水資源管理と河川保護の問題に取り組む国際機関や国際NGOとのパートナーシップを構築してきた。

中国の指導層は、政府だけでは環境問題（とくに水問題）をトップダウンの方式で解決することはできないことから、国際的な専門家を招くだけでなく、中国の環境NGOにもかなりの政治的スペースの拡大を許している。過去数年間、国家環境保護総局は環境法の策定および地方政府や企業の監視に人びとの関与の増大が必要であることを強調してきているが、それはそうしたボトムアップによる参加が政府による取り締まりや環境規制の執行における予算負担を軽減できるからである。中国政府が水汚染および水資源管理に関する法をさらに強化するために、水利委員会と関連法をトップダウンで改革するだけでなく、環境NGOと市民参加のより強力な推進に向けた改革が必要である。必要な改革としては、(1)NGOの登記がさらに容易になるような制度改革、(2)中国の企業や市民による地域のNGOへの寄付に対する免税措置の推進、NGOの国際組織への依存を断ち切ること、(3)環境政策の決定過程（環境アセスメント法など）やプロジェクトに関する情報への人びとやNGOのアクセスを増大させること、などが考えられる。

【コラム】水資源保護に取り組む中国の環境NGO

中国河川ネットワーク

中国河川ネットワークは中国の河川の保全に関心を持つ中国の環境NGOと個人のゆるやかな連合である。ネットワークは2004年に怒江ダム建設をめぐる論争が生じた最初の数ヶ月の間に、ダムに関する環境アセスメントの透明性を推進するための情報共有のプラットフォームとして設立された。このゆるやかなボランティア団体はNGOセクターにおける水問題に関する連絡組織として機能している。

贛江環境保護協会（江西省）

過去10年間、江西省に重度汚染型産業が多く立地するようになり、それにより贛江の水質が劇的に悪化した。贛江沿いの癌の発症率も増加するようになった。河川の急速な悪化に対して、2003年に関心の持つ市民や環境問題の専門家が贛江環境保護協会を設立した。協会は肖喬平を代表として4名のスタッフから成り、(1)水質調査の実施、(2)水資源保護に関する出版物の作成、(3)写真展および学校における講義の開催、(4)流域における環境保護の必要性に関するドキュメンタリー映画の撮影、を行っている。2003年7月から、協会のメンバーは贛江の中流域における水汚染問題に関する調査を行うためにバイクによるツアーを始め、15時間のビデオと何千もの写真を撮影した。地域および全国のメディアがこれらの写真の何枚かを使って、河川の水汚染問題と協会の活動について報道を行い、それによりこのNGOは大きな影響力を持つようになった。協会は2003年から河川のモニタリングによって、贛江の4つの支流に立地する企業の違法排水を暴露している。

緑色漢江（湖北省）

緑色漢江は2002年9月に登記された湖北省のNGOであり、また漢江の問題に取り組む最初の環境NGOである。緑色漢江は、人びとの教育や情報の普及を通して、漢江における環境的に持続可能な発展を推進することを目的としている。主な活動としては、漢江における環境ホットスポットに関する調査、地方政府機関に人びとの関心を伝えること、地域の汚染の監視、持続可能な発展に関する農村住民の教育などがある。緑色漢江は南水北調プロジェクトの建設により移転を余儀なくされる湖北省住民に対する補償の増額を主張している。

緑色江河（グリーン・リバー）（四川省）

緑色江河は、自然写真家の楊欣氏が、科学研究の推進を通して長江の脆弱な源流をいかに保護すればよいかについて政策立案者に情報提供を行いたいと考え、設立された。公式のNGO登記を行ったのは1999年であるが、四川省を拠点にした緑色江河としての活動は、1994年から二つの環境調査センターの運営を通して生態的に脆弱な長江源流地域を保護するために行われている。緑色江河の主なプロジェクトとしては、(1)地域の科学研究機関とジャーナリストと協力して、長江水源の水質に関する調査を行い、河川の健康状態に関する基礎的なデータを蓄積し、上流域の効果的な環境保護計画を策定することを支援すること、(2)地方政府による密漁パトロールの支援、(3)長江の生態系への脅威に関する農村住民およびツーリストの教育を行うボランティアの育成、などが行われてい

贛江中流にある製紙工場は汚水処理施設を導入せず、囲いを作ってその中で泡立つ廃水を貯め、希釈してから贛江に排出している（写真：肖喬平）。

る。緑色漢江は青蔵高速道路とラサと格尔木を結ぶ新たな鉄道の野生動植物に与える影響についてモニタリングを行い、報告書を出版している。新たな取り組みとして、緑色江河は眠江流域（長江の支流）のあるチベット族の村で生態学的に持続可能なツーリズムを推進するプログラムを行っている。村人達に対してエコツーリズムや環境保護に関する教育を行うほか、2006年には村で無水トイレと環境にやさしい固形廃棄物処理施設の建設を行う予定である。

緑色協会（黒竜江省）

緑色協会はハルビン工業大学の学生の環境活動家が結成した。1997年3月に地域の環境保護を目的として設立され、その影響力は徐々にハルビン市の外へと広がり、黒竜江省全体へと及んでいる。協会は何百、何千人もの人びとを巻き込んで廃電池のリサイクルに成功している。水資源の保護の分野については、協会はグローバル・グリーングラント基金の助成を得て、2003年4月に、松花江のいくつかの支流域において詳細な水質評価を行い、それがきっかけとなって無リン洗剤の使用を呼びかけている。また、黒竜江省における水質の深刻な悪化をまねいている土壌流出を防ぐために、2000年から毎年、何千もの地域の人々と植樹イベントを組織している。

グリーン・ウォーターシェッド（雲南省）

グリーン・ウォーターシェッドは、雲南省の瀾滄－メコン河流域の統合的流域管理に取り組む環境NGOである。グリーン・ウォーターシェッドは、中国西

南地域における参加型流域管理を支援するために必要な知識、技術、意思決定の方法を提供することを目的として、2002年に設立された。オクスファム・アメリカの支持のもと、グリーン・ウォーターシェドは、ラシ流域管理委員会を設立し、運営の支援を続けている。この委員会は広範囲の政府とコミュニティのステークホルダーの間における対話を進め、流域の開発と保護に関する方策を評価することを支援している。中国西南地域のダムをめぐる政策決定により広範囲の多様なステークホルダーの参加を推進するために、グリーン・ウォーターシェドは怒江流域の村民を率いて、万湾ダムおよび小湾ダムの村民を訪問する交流会を行った。この村と村の交流によって怒江流域の村民は、辺鄙な農村に建設されようとしているダムの決定要因について実地で見聞することができた。この交流ののち、グリーン・ウォーターシェドは、怒江流域の村民が中国の報道機関を通して意見を表明する機会を設けた。こうした報告によって怒江のダムの決定過程で多くの草の根の声に耳が傾けられる機会を与えられたのである。

淮河衛士（河南省）

淮河衛士は、河南省周口市沈丘県の民政局に2003年に登記されたNGOである。代表の霍岱珊氏は、フォト・ジャーナリストであり、写真展を通して、ひどく汚染された淮河による健康被害および生態系破壊の深刻さを訴える活動を行っている。霍氏が流域村民の健康調査を行ったところ、河川の水汚染に起因するとみられる癌の発症率が異常に高いことを発見した。汚染された河川や水路沿いにおいて彼が行った多くの健康調査から、癌患者の異常な高率が見られる村は100以上あると見込まれている。霍氏は民間基金から助成を得て水の濾過装置や薬品をいくつかの村に提供している。CCTVや他の報道機関が彼の健康調査について報道を行い、こうした癌の村における活動を支援している。そうした報道が後押しして、地方政府は村々に安全で清浄な飲み水を提供するための深井戸の掘削を行うようになっている。

三江源生態保護協会（青海省）

このNGOの名前となっている三江源は、青海平原に位置する長江、黄河、メコン河の3つの河川の水源である。この地域の美しさは有名であるが、過去10年間にわたり、青海平原の生態環境は、地球温暖化と無秩序な開発によって急速に破壊されてきた。水源地域の20%以上の動植物種が危機にさらされている。三江源生態保護協会は2001年に地元のチベット族によって、環境教育、公衆参加活動、コミュニティによる土地利用計画の策定によって三江源流域を保護するために設立された。協会は、地方政府と研究者と協力して、専門家の諮問委員会を設立し、三江源の環境保護活動を監督している。協会はまた「積雪地域および大河川の環境教育移動学校プロジェクト」を実施し、中国における多くの人びとが三江源の脅威を知り、伝統的なチベット族による環境保護活動の実践を学ぶ機会を提供している。

第3章 中国の河川保護に関する国際協力の潮流

　過去20年間、多くの国際組織が国務院、全国人民代表大会、国家環境保護総局、水利部、および他の政府部門と共同で新たな環境政策、規制およびパイロットプロジェクトの推進に取り組んできた。1990年代における水の汚染と不足から生起する健康への脅威と紛争の増加に伴い、中国政府はこのセクターにおけるより多くの国際援助を求めている。本章では中国におけるそうした国際援助について概観する。

　　世界銀行は数多くの水資源保護プロジェクトに従事しており、小流域における用水戸協会やタリム湖と海河流域で流域ガバナンス制度の能力改善を直接ねらった二つの注目すべき取り組みを行っている。

　2003年より、アジア開発銀行（ADB）は黄河において、区域を越えた環境管理プロジェクトという研究を始めている。この研究は、黄河を守るために、法規、資金調達、管理、社会変動に焦点をあてた部門横断的なものである。

　英国開発援助庁（DFID)は、他の国際機関と協力して、中国政府が2003年の水法改正において提案されている水セクターの改革を進めるためのプログラム－水利用者の参加、水資源管理へのより統合的なアプローチ、水土保持のための新たなアプローチ、飲み水と衛生設備へのアクセスの増加など－を実施することを支援している。

　5年以上にわたり、EU北京事務所は、EU、日本、世界銀行の協力を得て、遼寧省政府とこの水不足が深刻な流域における持続可能な流域管理を促進する広範なプロジェクトを実施している。

　過去数年間、スウェーデン国際開発協力庁（Sida）は次の領域において多数の水関連の協力プロジェクトを支援してきた。(1)工業におけるより水利用効率の高い技術、(2)内モンゴルにおける湖の再生に向けた包括的なアクションプランの作成、(3)農業の水利用効率の向上、(4)下水処理場のよりよい管理のための能力向上、(5)生態的衛生システムの開発。

日本政府による中国における水関連事業

　1990年代中頃より、中国に対する日本のODAは非常に大きなシェアで環境プロジェクトに向けられてきた。このうち多くのプロジェクトが、水問題、とりわけ下水処理場、上水供給施設、大規模灌漑地域の節水、流域環境改善に焦点があてられている。2004年、日本政府は中国において、植林、砂漠化対策、流域管理など水資源の管理や保全に重点を置くとした。さらに日本はこれまでの対中援助を基礎として、水汚染や生態系の保全の問題に取り組む予定である。日本のODAによる中国の環境保護への貢献は大きいものの、日本政府は現在中国政府と、中国初のオリンピックが北京で開催される2008年までに、円借款の供与を停止する方向で調整を行っている。もし多額の円借款が停止されるとなれば、中国に対する日本のODAはインフラプロジェクトよりも制度改革支援や人材開発事業に重点が置かれることになるであろう。

国際協力銀行（JBIC）

多くの開発途上国に円借款を供与している国際協力銀行(JBIC)は、中国で、西部地域における環境、人的資源開発、および貧困削減の3つの領域に重点を置いている。1979年より、JBIC（前身はOECF）は中国に多額の借款事業を行ってきており、過去5年だけでも円借款は毎年平均15億ドルに及んでいる。JBICは流域管理を直接支援するための特別のプロジェクトを行っていないが、多くの水に関連するプロジェクトを実施している。

国際協力機構（JICA）

国際協力機構(JICA)は日本の専門家を中国に派遣、あるいは中国人関係者を日本で研修するなどして以下のような水関連の技術協力を行っている。(1)2,000人以上の中央・地方の水利部門の官僚を対象とした水資源開発に携わる人材養成事業、(2)大規模灌漑区における節水灌漑モデル計画、(3)太湖における水環境再生パイロット事業、(4)四川省森林造成モデル計画。これまで日本の水関連協力事業の多くは水管理のハードウェアに関する技術移転に焦点をあてたものが多かったが、最近のJICAのプロジェクトには、上記紹介したように水セクターにおける人的資源や政策の強化など、水管理のソフトウェア的要素のより多い取り組みも見られる。さらに、JICAは中国の水利権整備プロジェクトをたちあげ、国土交通省の専門家や日本国内の学識経験者からの支援を得て、遼寧省の太子河流域のケーススタディを実施している。また、大規模灌漑区の節水灌漑モデル計画事業の一部として、日本農業土木総合研究所は土地改良区を含む日本の経験を紹介する技術交流プログラムを実施している。日本における土地改良区の経験は40年以上に及び、参加型灌漑管理の成功事例として中国にとっても大いに参考になる。

アメリカ政府による中国における水関連事業

日本と対照的に、アメリカの政府機関は環境プロジェクトのために中国政府に対して借款供与も無償資金の提供も行っていない。中国への直接援助に関する議会の制約のために、中国において、20近いアメリカの政府機関が公的な開発援助ではなく内部の予算を使って、100以上の環境・エネルギー関連事業を行っている。

2000年以来、米国農業省(USDA)と環境保護庁（EPA）は山東省の黄河下流域において水質モニタリング、下水再利用、流域管理に関するデモンストレーションプロジェクトを行ってきた。

USDAの経済調査サービスは、中国科学院、水利部、オーストラリア農業資源経済局、カルフォルニア大学（デービス）と協力して、中国における水資源と農産物に関する調査を行っている。2003年より、黄河流域モデルに関するデータを収集し、2005年には流域における水取引と環境流量に関する初歩的なシナリオを作成した。

2006年、USDAの経済調査サービスは水利部の研究者と共同で、黄河流域モデルの水文セクターへの影響を取り入れた灌漑管理改革と節水灌漑技術の適用に関するインプリケーションをさらに研究するための新たな取り組みを提案している。

2006年、EPAは海河流域において、中国のサステイナブル・シティのための清浄な水プログラムを完成予定である。この水質に重点を置いたプロジェクトは、天津市環境保護局、国家環境保護総局、水利部、海河水利委員会、地球環境基金(GEF)、ADBと共同で行われている。このプロジェクトの目的は、安全な飲み水と衛生設備への人びとのアクセスを向上させ、天津市に近い海河流域の流域管理を促進することである。

低い廃水処理率（全国平均40%）のために都市が中国河川の主な汚染源となっている（写真：肖喬平）。

国家環境保護総局とEPAの間で2003年に締結された協定をもとに、新たな水汚染防止事業が始まっており、中国における排水権取引に関するパイロットプロジェクトを実施するための覚え書きが交わされている。

中国における水問題に取り組む国際環境NGOと研究組織

過去数年間、国際NGOは流域保護と管理の分野においてより多くの事業を行うようになった。アメリカの環境NGOのいくつか－WWF、コンサーベイション・インターナショナル(CI)、ザ・ネイチャー・コンサーバンシー(TNC)など－は中国の河川流域保護に積極的であり、最重点あるいは2番目の重点事業として取り組んでいる。日本の環境NGO－ラムサールセンター(RCJ)、メコン・ウォッチ、日本環境会議など－や研究機関もまた、中国の流域をテーマにしたスタディツアー、共同研究、国際会議やワークショップを積極的に実施している。これら国際NGOや研究機関の水関連プロジェクトは中央、省、地方政府機関、研究機関、市民社会団体を集め、ネットワークを構築している。すなわち、こうしたプロジェクトは中国における水資源保護をめぐるコミュニケーションの新たな体制を形成し、ステークホルダーの参加を増大させている。

結論

中国における水問題の重大性と政府の海外援助への開放的な姿勢により、全国の水資源管理と汚染防止において国際組織の参入が拡大している。過去数年間、国際組織は小さなプロジェクト中心の取り組み（例えば、廃水処理施設など）から、流域あるいは国家レベルの政策に焦点をあてた意欲的な取り組み（例えば、水利委員会の改革や水利権整備など）へと移行してきている。中国における国際的な水プロジェクトの数と範囲は増大・拡大しているが、この事業を実施する組織はそれぞれのプロジェクトの成功あるいは学ぶべき教訓をほとんど共有していない。上記で紹介したような多様な国際的な河川関連事業は、中国の流域ガバナンスに関する日米協力の可能な選択肢について知見を提示してくれる。

第4章　中国における流域ガバナンスに関する日米協力の可能性

　過去15年間、中国政府による水資源政策の強化と主要河川・湖沼の浄化に向けた意欲的な目標とキャンペーンにもかかわらず、中国の水質―とりわけ河川―は著しく悪化している。水法の改正および国際協力によって中国において統合的水資源管理の概念が普及した。中国の水問題を緩和するため、とりわけ河川流域を保護するための政策の改正と国際協力には、創造的発想と、グローバル、リージョナル、ナショナル、サブ・ナショナルの各レベルにおける組織の環境問題に関する専門家および実務家との対話が必要である。今後10年間、中国における持続可能な流域の開発に向けた支援は非常に重要であり、この重要な領域においてとくに日米両国は可能な限り協力―あるいは少なくとも努力―し、またそれぞれの経験と技術の共有を図るべきである。

　日米両国の政府、NGO、研究機関は中国の環境保護問題、とくに水問題についての援助と研究に積極的である。しかし、情報の共有がほとんどなされておらず、また日米両国における公式の共同事業もない。経済不況、地政学的な優先課題の変化、および最近の大規模な自然災害によって、日米は国際開発援助をいくらか削減することを迫られている。それゆえ、国際環境協力における情報共有と共同事業によって、両国が縮小している開発援助財政の効果を向上させ、同時に中国および他の開発途上国における不必要なプロジェクトへの投資を避けることができる。

　以下、私たちはまず、日米両国政府が国際協力プログラムにおいていかにして水問題を優先課題とするようになったかについて議論する。次に、中国の河川を保護するために、両国における政府、NGO、研究機関の間で、統合的流域管理の3つの重要なテーマ、(1)流域管理機構、(2)資金調達メカニズム、(3)公衆参加、をめぐる協力可能な分野を示す。

国際協力における水問題の優先
　日米両国はそれぞれの国際協力プログラムにおいて水問題に高い優先順位を与えており、しばしば開発途上国における貧困削減あるいは都市開発といったより広い分野のひとつとして取り組みがなされている。2003年、日本で開催された世界水フォーラムにおいて、開発途上国の水問題に関する国際協力を拡大することの必要性が提言された。この提言の精神にもとづき、世界水フォーラムにおいて、水資源機構とアジア開発銀行はアジア流域機構ネットワーク（NARBO）のプロジェクトを開始した。水資源の開発および保全に関する日本の経験を生かして、NARBOはアジア諸国における流域レベルでの統合的水資源管理（IWRM）を促進することを目的として、政策提言、トレーニング、技術的アドバイス、地域協力を行っている。

　米国開発援助庁（USAID）は、世界における水資源の保護と環境にやさしい開発を最優先課題としている。世界各国の水セクターにおけるUSAIDのプロジェクトと投資は、安全で十分な水供給と衛生設備へのアクセスの改善、灌漑技術の

改善、水系生態系の保護、水資源管理の制度的能力の強化に重点を置いている。2003年から2005年の間、USAIDは170億ドル以上の資金によって、76カ国の開発途上国における淡水および沿海の持続可能な資源管理を改善し、2,800万人近い人びとの衛生設備へのアクセスを改善し、流域ガバナンスに取り組む約3,400の団体を流域レベルの統合的水資源管理の政策決定に動員した。USAIDは、ポール・サイモン上院議員が提案し、2005年11月30日に署名された救貧法にもとづき水関連事業を拡張した。この新たな法律の目的は、安全で十分な飲み水と衛生設備および持続可能な水資源管理を、アメリカの外交政策の要とすることである。

それぞれの独立した国際協力に加えて、日米両国は共同で水関連プログラムを強化する道を開いている。2002年の持続可能な開発に関する世界サミットにおいて、日米両国政府は新たな水協力イニシアティブ（日米水パートナーシップ）を立ち上げ、両国が共同あるいは並行して開発途上国の水関連プロジェクトを行うことに合意した。USAIDと国際協力銀行（JBIC）は、フィリピン、インドネシア、ジャマイカ、インドの4カ国において、水関連の資金援助プロジェクト実施の中心的な役割を果たしている。例えばフィリピンにおいて、2つのパイロットプロジェクトが進行中である。そのうちひとつのプロジェクトでは、地方自治体水関連ローン・ファイナンシング・ファシリティが、JBICが支援する信用供与とUSAIDの開発信用制度が支援する民間融資を利用する予定である。また、2005年初めには、フィリピン水リボルビング・ファンドのフィージリビリティ・スタディが完了し、2007年初めには開始される計画である。同様の資金調達関連プログラムは他の3つのパイロット諸国で計画中である。中国はこの協力プログラムの対象ではないが、こうした水関連資金調達に関する日米協力が実施されれば、その恩恵を受けることができるであろう。

協力の可能な分野

中国の水問題解決の要請は大きくまた複雑であるが、私たちは、管理機構、資金調達、公衆参加に重点を置いた統合的流域管理の概念を援用して、日米両国（および他の諸国）が中国における水関連協力の可能な分野を考える上で触媒となるよう、様々なアイデアがつまったたくさんの「引き出し」を提供したい。相互協力の可能性は日米両国政府のみならず、両国のNGOや研究機関にもある。

法制度改革

中国における統合的流域管理の概念を普及させるための法的および制度的改革を促進するうえで、日米両国政府は共同で、水利委員会が設置されている七大河川流域のいずれかのひとつの小流域（支流、湖沼、あるいは河口域）で、小さなパイロットプログラムを立ち上げることができるであろう。パイロットプログラムは小規模な制度改革－水利権、用水戸協会、価格政策など－に焦点をあてるとよい。中国発展・環境国際協力委員会（CCICED）の統合的流域管理タスクフォースが提案している意欲的なパイロットプロジェクトとして、支流、湖沼、あるいは河口域レベルにおいて、省政府、地方行政、および各ステークホルダーの代表からなる管理委員会を創設することが考えられるであろう。こうした地域レベルの流域管理委員会は、流域の計画および目標を設定し、ステークホルダーの参加や流域の保護を進めるための経済的インセンティブの試みについてモニタリングを行うことができよう。

このようにボトムアップで水利委員会の改革を行おうとする現場型パイロット事業は、水利委員会のメンバーが、数ヶ月間、日米の河川流域委員会において実施されているプロジェクトの視察を行う交流活動により強化することが可能で

長江の第一湾曲部にあたる金沙江ではダム建設に大きな注目が集まっている。そのダムは長江の支流ではな
く、長江の本流に建設されようとしている。このダムは虎跳峡まで続く金沙江に計画されている12のダムの最
初のものである。これら長江上流部でのダム建設計画の目的は、水力発電のみならず、三峡ダムにおける土砂
の過剰堆積を防ぐことである（写真：馬軍）。

あろう。流域におけるあらゆるステークホルダーの参加メカニズムを有している
河川流域委員会を訪問することによって、中国の河川管理者は、どのようにそれ
を取り込むかということだけでなく、どのように水紛争を解決するかについての
知見を得ることができる。中国において、政府は人間のニーズと生態系保全のバ
ランスをとる開発ではなく、もっぱら経済開発のために河川の管理を行っている
ことから、国内および国際的な水紛争を解決するという課題はますます増大して
いる。他方、日米および他の先進諸国や国際機関は河川流量の生態学的価値をま
すます重視している。河川流域については日米でもそれぞれなお論争がなくはな
いが、両国ともに河川流量の生態学的価値を強調する法制度を策定し、参加と紛
争解決のための手続が用意されている。

　アメリカの流域委員会のうち、4つの州と連邦政府をメンバーとしているデラ
ウェア流域委員会は研究に値しよう。委員会は、1961年に設立されて以来、州間
の絶え間ない紛争を解決してきただけでなく、政府、市民、NGOを効果的に動
員するフォーラムとして機能し、水不足と水汚染問題を解決してきた。省政府を
メンバーとせず、十分な権限を持たず、包括的な取り組みのできない中国の水利
委員会と対照的に、デラウェア流域委員会は、十分な規制権限と多様なステーク
ホルダーを動員する能力を付与されれば、委員会がいかに河川流域のよりよいガ
バナンスに到達できるかについて、興味深いモデルを提供してくれる。

　日本の河川流域委員会もまた中国の関係機関にとって貴重な事例を提供して
くれる。というのは、両国ともに河川の管理においてきわめて中央集権的なシス
テムを有しているからである。1997年、日本で河川法が改正され、多くの河川や
湖沼で流域委員会が設立されるようになった。日本においてこうした委員会は比
較的新しい試みであるが、ステークホルダーの動員、流域の開発と環境に関する
微妙な問題についての合意形成をめぐってかなりの経験を蓄積しつつある。例え
ば、淀川水系流域委員会は4年半にわたり400回以上の流域管理計画に関する会合

を公開で開催しているユニークな諮問組織である。この委員会は国土交通省地方建設局の主導で設置されたが、その運営は実質上、地方建設局ではなく、学者、コミュニティ、NGOの代表からなる委員の協議で行われている。委員会の事務局は民間企業が請け負っている。公開の議論にもとづく合意形成のプロセスはゆっくりとしたものであり、まだ流域計画草案について議論を行っている。しかし、その計画がいったん実施されれば反対は少なく、紛争が生じても比較的容易に解決されるであろう。

また、CCICEDの統合的流域管理タスクフォースの報告書による提案として、上記のような現場型実践の積み上げに加えて、中国における水のガバナンスに関する国家レベルの制度的かつ法的な変革が示唆されている。例えば、国際組織は、中央政府と連携して、水利委員会の矛盾を軽減し、責務を明らかにするために、流域管理および水汚染防止に関する法制度の評価・改正を支援することが可能である。ひとつの法改正支援の方法として、議会間の交流が考えられる。例えば、全国人民代表大会環境・資源委員会の委員が日米のカウンターパートと会い、両国が河川および水資源の保護との関連で策定したより効果的な法律について学ぶことができるであろう。

また、水に関係する行政部門の縦割りの弊害を緩和するために、統合的流域管理タスクフォースが提案している別のハイレベルの取り組みとして、国家発展・改革委員会、水利部、国家環境保護総局を含む国家レベルの統合的流域管理委員会の創設がある。この委員会は全国的に統合的流域管理を推進するための法改正や新法制定を監督することができる。

他の多くの国々と同様、中国において、淮河における水汚染事故や黄河における断流といった水危機が、河川保護のための政府機関の協調を引き出す契機となっている。しかし、中国の河川における水危機への対応としての協調は、しばしば河川管理権限のさらなる中央集権化か、あまり効果のないキャンペーンをまねくだけで、持続可能な水のガバナンスに向けた組織統合の理想的な要因とはなっていない。政府部門間協調を推進するためのインセンティブメカニズムを構築することは、中国において強力な統合的流域管理を構築する上で重要である。そのような協調を引き出すためには、以下の二つの重要な研究分野が考えられる。(1)異なる専門行政組織（例えば、上水、下水、洪水防御に関する行政）間の協調がいかに、またどのような時に行えば、社会的な効用を生むかを明らかにすること、(2)どのように十分な政治的意志を動員して複数の専門行政組織を協調させるかを検討すること、である。情報およびデータの共有こそが、水資源保護政策の計画と実施の改善だけでなく、水問題の解決の費用と時間を節約し、結果としてあらゆるステークホルダーを利するよう、中国がとるべき最初の現実的なステップであろう。

新たな資金調達メカニズムとインセンティブの活用

「誰が利益を得、誰が費用を負担するのか」は持続可能な流域管理をめぐる経済学の主な論点である。この問題に答えるために、中国の政策立案者は市場メカニズムを基礎とした手段を、環境規制や環境保全の促進の新たな手段として導入することに熱心なようである。しかし、中国のあいまいな水利権や弱い法制度は水に関する市場メカニズムの体系的な適用の妨げとなっている。

中国における水利権は非常に複雑でかつ微妙な問題である。しかし、米国農業省（USDA）と日本の国際協力機構(JICA)はそれぞれ中国における水利権の設定に関するプロジェクトを実施している。公式的なパートナーシップは実現が難しくとも、USDAとJICAがパイロットプロジェクトから得る知見とこの分野にお

ける研究について、より広く共有、普及するべきである。そのような情報共有はこの分野における共同事業の可能性を探る上で役に立つであろう。

　上水システムおよび下水処理場の建設・運営に関する費用をカバーするために、中国において水の価格を引き上げることが求められている。中国における水資源の保全のための費用負担の最適なモデルはまだなく、中国の持続可能な流域ガバナンスの緊急かつ困難な課題となっている。日米両国の政府およびNGOは、中国政府と協力して、河川や湿地の水資源の保全を推進するための資金調達メカニズムを開拓するための事業を行うことが可能である。中国政府の関心は高いものの、水資源の保全を推進するための資金調達（例えば、環境資源の利用に対する費用負担制度、環境税、リボルビング・ファンド、市債など）や市場メカニズムの活用（例えば、水利権取引や水基金など）に焦点をあてた中国国内あるいは国際的な取り組みは非常に少ない。こうした水関連の資金調達の課題について、以下に日米中の協力が可能な分野を挙げておく。

　(1) 環境資源の利用に対する費用負担制度*(payment for environmental services)*に関するパイロット事業: 下流の水利用者が上流域の環境保護についていかに補償（費用負担）するかは、多くの国々が直面している課題である。アメリカでは、USAIDが、開発途上国における河川の環境資源に対する費用負担制度に関する多くのパイロットプロジェクトを支援しており、この制度に関するひとつの成功モデルといえるであろう。もうひとつのモデルとして、日本の都道府県により、上流域における水源および森林の保護を推進することを目的として導入・実施されている新たな税制があげられる。このようなグリーン税制への取り組みは日本においてまだ2年ほどの経験しかないため、河川保護の効果を評価するのは時期尚早である。しかし、中国の政策立案者および河川管理者がこのシステムを研究することで、上流と下流のステークホルダーのパートナーシップをもとに、水源を保護するための経済的インセンティブの手段を、いかにして導入するのかについて学ぶことができよう。中国にはこうした方策の前例として、政府が森林資源の保護のために適用している補償制度がある（主に、費用徴収、補助金、税金、罰則など）。例えば、木材伐採は中国西南地域の多くで禁止されており、農民が傾斜地の農地を森林に転換できるよう所得補填と補助金の支出を行って植林を奨励している。しかし、この森林保護プログラムは、政府によるものであり、市場メカニズムによるものではない。

　(2) 水質・水量の保護のための用水戸協会の普及: 　世界銀行は中国において2,000近い用水戸協会の設立を支援しているが、農村地域の多くの地方水利部門は、十分なサービスの提供や十分な水費の設定に際して深刻な困難に直面している。海外における他の水利用組織の成功例、とりわけ水汚染防止を促進するような例の検討が必要なことは明らかである。例えば、オランダにおいて、地域のステークホルダーにより構成されている水管理組合が、メンバーによる排水課徴金の料率を設定する上で重要な役割を果たしている。

　(3) 水資源の保全と水汚染の防止のためのリボルビング・ファンドの創設: 1987年に米国議会は清浄水法を改正した際に、連邦水清浄リボルビング・ファンド（CWSRF）プログラムを創設した。CWSRFプログラムによって、伝統的な地方自治体の下水処理プロジェクトに加えて、非点源汚染、流域の保護と再生、湿地管理プロジェクトなど、様々な水質プロジェクトに資金の提供が可能となっている。そうしたファンドは中国におけるいくつかのパイロットプロジェクトの目標でもあった。例えば、2004年、長江中流域の湿地再生プロジェクトの一部とし

て、WWF中国は、湿地の再生によって土地を失った秋湖村の農民が代替的生計手段を開拓できるよう、リボルビング・ファンド制度を立ち上げた。最初に提供する資金によって、農民達による竹の苗床、持続可能な漁業、エコツーリズムのベンチャー、水耕栽培を支援した。最初の回転資金において、104世帯がローンを返済し、それを原資として他の農民達が新たなローンを組むことが可能となった。

(4) 下水処理場への資金提供のための市債（地方債券制度）の活用: 水汚染防止において、中央政府は下水処理場の費用をカバーするための公的な政策を有しておらず、地方政府はしばしば地域経済を妨げるとして下水処理への投資を嫌煙している。ひとつの解決方法として、それは中国における法制度および資金調達における大きな変革を要するものではあるが、下水処理場に対する市債の創設が考えられる。世界銀行と米国貿易開発局は山東省に下水処理場に対する市債のパイロットプロジェクトを実施・完了した。投資家が地方債券市場に参入し、市債のリスクを最小限にするメカニズムと法制度を構築する上で現行法における必要な調整事項を明らかにするために、同様のパイロットプロジェクトを、地方政府と国家発展・改革委員会の協力を得て、他の都市で行うことが可能であろう。

(5) 小規模な水取引の実験: 第2章で述べたように、中国において、合法的ではないが、いくつかの水利権取引が行われているが、日米両国は、現行の水利権事業をもとに、地方レベルにおける水取引に関する制度構築を支援することが可能であろう。

(6) より広範な費用と便益を環境アセスメントおよび計画に組み込む: 資金調達に関する国際的な様々な取り組みを、CCICEDの統合的流域管理タスクフォースの提言に従い、ある河川流域におけるパイロットプロジェクトとして実施することが可能であろう。理想的には、水利委員会による河川の開発と計画の決定が、経済的コストのみならず、社会的、環境的コストの評価基準を含んだ環境アセスメントをもとに行われるべきである。日米中の共同チームにより、河川管理に対する真の環境的、生態的、社会的影響評価を組み込むための障害と解決の可能性を明らかにするような研究やパイロットプロジェクトを実施することが可能であろう。

ステークホルダーの参加のさらなる拡大

1990年代中頃から、中国政府は、政府と市民の間の協力を促進する国際協力プロジェクトによるだけではなく、改革期における社会のさらなる開放性を認める政治的変化によって、環境分野において公衆参加を積極的に奨励するようになってきている。環境保護における市民参加の拡大と市民と国家の関係の顕著な変化を促してきた改革として、いくらか開かれてきた報道機関、個人によるNGOの設立を認める制度、被害を受けた個人が訴訟を起こす権利、環境アセスメントにおけるパブリックコメントの要求、情報へのアクセスの漸進的な増加などがある。こうした動向は、流域管理における公衆参加の拡大において国際的な取り組みに希望を与え、またその機会を開くものである。第2章と第3章において、国内および国際NGOのいくつかの事例とこの分野における二国間援助を紹介したが、そうしたプロジェクトがさらに必要とされている。

河川管理における公衆参加の基本的な要件として、流域のステークホルダーがすべての情報にアクセスでき、また統合的流域管理計画、環境アセスメント、および流域における管理の決定に意見を表明できることをあげることができる。

日米両国の政府、NGO、研究機関には水資源管理および汚染防止の分野において、ステークホルダーの参加の拡大を後押しするような協力に関する多くの機会が開かれている

(1)流域レベルのフォーラム: CCICED統合的流域管理タスクフォースは、各大河川流域において、省間および政府、NGO、研究者といったステークホルダー間の対話と合意形成のためのプラットフォームとして機能するような、開発と保全に関するフォーラムの創設を提案している。

(2)支流域あるいは地域単位における企業の社会的責任(CSR)を促進するフォーラム: 企業が主な汚染源であるが、企業や地方政府にコミュニティと協働するよう働きかけるのは容易ではない。しかし、水資源の保護のためのCSRがいかに企業にとって利益となるかについて、支流域あるいは都市における企業や他のステークホルダーが理解を深めるための取り組みを国際組織が行うことは可能であろう。水汚染防止のCSRの例としては、政府による排出

贛江の風景（写真：肖喬平）。

基準よりさらに高い水準への自発的な取り組みや人びとへの排出情報の開示、グリーン・サプライ・チェーンの義務づけ、NGO－企業間のパートナーシップの構築、水汚染物質排出権取引のパイロットプロジェクトへの自発的な参加、透明性の高いエネルギー管理システムの構築などがあげられる。

(3)流域間の交流: 中国において流域管理に携わる政府官僚やNGOが、日本やアメリカを訪問し、人びとがいかにして河川や流域の管理に参加しているかについて学ぶことが可能である。

(4)流域管理決定過程における公聴会: 統合的流域管理のパイロットプロジェクトにおいて、流域管理計画とその実施の際に、コミュニティメンバーが恒常的に参加する機会を設けることは重要なメカニズムのひとつとなろう。現在、中国における多くの公聴会は、ほとんど決定がなされた後に、人びとが単にコメントを表明するための会合となっている。2005年11月に、国家環境保護総局は、環境アセスメントにおける公衆参加を前進させるための実施細則を策定するうえでのアドバイスを国際社会に求めた。この新たな規定が策定されれば、総局はトレーニングを実施するための支援を必要とするであろう。

(5)汚染被害者への法律援助: 水汚染事件において市民が法廷にアクセスできるよう支援することは、地方政府や企業に現行の水汚染防止処理関連法の実施を迫る上で大きな役割を果たしうる。現在、汚染被害者を支援する事業に従事している中国のNGOは1団体のみであり、このことはこうした法律援助の取り組みが中国において不足していることを意味している。

(6)中国NGOのトレーニング: 中国の環境NGOの大部分は、国際NGO、海外の財団、他の政府機関からの支援により、能力の向上と影響力の増大を図ってきた。こうした外部からの支援は欠かせず、とりわけ水問題に関してさらなる支援

が可能であろう。例えば、流域問題について二国間および多国間援助による多くの取り組みがあるが、NGOを巻き込んだものはきわめて少ない。そうしたプロジェクトへのNGOの巻き込みは、NGOを流域管理における正規の参加者とするためにきわめて重要であろう。

(7)スチュワードシップ(*stewardship*)の醸成：政府と人びとの間でパートナーシップを促進し、河川を共同で管理することは重要であるが、さらに重要な変革は、市民が水資源およびその周辺の土地の管理受託者(stewards)となれるよう、市民をエンパワーメントする（力を与える）ことであろう。流域資源の管理により多くの市民が参加することによって、政府の河川保護に要する費用をより節約できるであろう。日本やインドの湿地において活動するラムサールセンターは、市民の政府機関とのパートナーシップを促進するのに成功しただけでなく、いかにして沿岸のラグーン生態系を再生するかということと同時に、いかに自らの生活改善を図るかについて、市民が主たる政策決定者となるようエンパワーメントしている。水資源管理へのこのスチュワードシップ・アプローチは、規制アプローチを補完しつつ、生態系における重要な財とサービスを流域単位で長期にわたり維持するために、長期的なビジョンと持続的な行動を獲得する上で最も有望な方法である。

結論

急速な経済成長を遂げる「世界の工場」として中国は、製品の輸出入の両面で世界市場に大きな影響を及ぼしており、さらにその影響は今後10年間ますます大きくなるであろう。環境保護と省エネルギーにおいて中国政府が現在とる選択は将来にわたりグローバルな影響を持つであろう。中国は、持続可能な発展のモデルへの転換に向け、環境保護と省エネルギーに関する法制度の形成において顕著な前進を見せている。しかし、それら法制度の実施と執行は一様でなく、とりわけ水問題についてそれが顕著である。

中国が水セクターにおいて直面している課題には困難が多い。他方、中国政府は、水資源管理の国内改革に向けて開放的な姿勢をとっており、また水問題を緩和するために海外の事例に関心を持っている。このことは、日本、アメリカ、および他の諸国にとって、中国を支援する重要な機会が訪れていることを意味している。日米両国が中国に対し、流域管理に関する異なる背景と経験を提供することで、中国は河川の保護に関する多くの選択肢とアイデアを引き出すことが可能となる。そうした協力は他の諸国における日米の新たな水パートナーシップを促すことにもなろう。すなわち、太平洋を越えて、中国の持続可能な流域ガバナンスを促進するために日米が協力することで、中国だけではなく、世界における水の安全保障に貢献することができるであろう。

前言与致谢

　这份关于中国水资源挑战的报告源起于美国华盛顿特区威尔逊中心中国环境论坛的吴岚 (Jennifer Turner)与日本贸易振兴机构亚洲经济研究所的大冢健司 (Kenji Otsuka) 的共同合作项目—建立美日的水资源伙伴: 促进中国可持续的流域治理, 而且这个项目的完成是由日本国际交流基金会全球伙伴中心的纽约办公室所慷慨赞助在这个项目中, 这份横跨三国的研究团队成员所完成的研究报告(建立美日的水资源伙伴: 促进中国可持续的流域管理)出版在2005年3月亚洲经济研究所焦点调查的第二十八期。2005年11月亚研世界潮流 (Ajiken Word Trends)(日本贸易振兴机构亚洲经济研究所的日文出版刊物, 讨论发展中国家的前景)的特刊出版了研究团队日本研究组员的报告, 是关于中国可持续流域治理的国际合作。这份报告包含亚洲经济研究所焦点调查中的一些研究, 但主要还是更新的资料。中文和日文的印刷版概括了英文原文的主要观点。全文的日文和中文翻译可以在中国环境论坛的出版刊物一栏下找到。网址为www.wilsoncenter.org/cef。

　在出版此报告的过程中, 我们需要感谢很多个人及组织机构的大力协助。首先要感谢的是所有热心参与中国, 日本与美国流域治理研究考察团的成员, 这十位成员包括: Carol Collier, 藤田香, 胡勘平, 片冈直树, 中村玲子, Richard Volk, 王亚华, Gary Wolff, 山田七绘, 于晓刚。这十位成员利用考察团设计发展个人的研究报告, 探讨中国如何向日本和美国学习三个关键领域(流域治理机构, 财政和公共参与)的经验, 进行流域综合管理。另外一位关键成员是 Timothy Hildebrandt, 在他攻读博士学位前, 曾经是中国环境论坛的项目助理, 他为研究考察团的观念与构成贡献良多。我们也想要感谢荣誉成员—中山干康—他为我们的第一份出版物贡献了他对于中国境内跨国界河流的宝贵研究。此外, 琵琶湖研究所的中村正久与京都大学的 北野尚宏, 在2005年10月7日于东京举办的国际研讨会中, 发表他们的研究, 并提供我们许多宝贵的建议。

　在研究考察团的过程中, 有无以计数的朋友慷慨地与我们分享如何有效治理河流流域的经验与专业。依照英文字母顺序, 我们诚挚地想向以下组织的工作人员致上谢意, 谢谢大家对于我们研究团队的协助: Asaza基金, 奇瑟比克湾基金会, 奇瑟比克湾项目, 中国环境与发展国际合作委员会, 中国环境与可持续发展资料研究中心, 中国水利部, 国家保护组织的北京办公室, 达拉瓦流域委员会, 瑞典大使馆在北京, 欧盟的北京办公室, 中国绿色时报, 中国绿色评论, 绿色江河, 绿色流域, 德国技术合作公司, 天津的海河流域委员会, 国际建设技术协会, 日本贸易振兴机构亚洲经济研究所, 波多玛克河跨州委员会, 日本国际协力银行, 日本国际协力机构, 日本水资源机构, 神奈川县的税务改革办公室, 马里兰自然资源部, 桃山学院大学, 纽约区域计划协会, 中国人民大学, 太平洋研究所, 日本拉姆萨中心, 天津环保局, 东京经济大学, 东京都政府港湾局地方办公室, 中国清华大学, 日本筑波大学, 英国国际发展部, 东京大学, 美国国际发展部, 美国工兵团, 美国环保局, 国际湿地联盟的北京办公室, 威尔逊中心, 世界银行的北京办公

硫和工厂排放的其它流走物质——如长江支流大渡河岸上的这些工厂——恶化了侵蚀、污水以及垃圾等造成的长江水质问题。照片提供：杨欣

室，世界渔业中心，以及世界自然基金会在中国的办公室。

此外，我们还需要感谢很多人为我们的报告提供大量宝贵的建议或协助阅读报告的初稿，帮助我们强化或澄清报告的论点：史丹佛大学的Baruch Boxer，瑞典大使馆北京办公室的，美国农业部的 Bryan Lomar，藤田香(桃山学院大学)，Jim Nickum (东京女学馆大学)，Richard Volk (USAID)，中国清华大学的王亚华，东京经济大学的片冈直树，温波 (Pacific Environment)，日本贸易振兴机构亚洲经济研究所的山田七绘，中国水危机的作者马军，香港中文大学的 吴逢时，森尚树(中日友好环保中心项目)，石渡干夫(日本国际协力机构)。我们也需要感谢中国环境论坛的四位研究助理，包括 Linden Ellis, Charlotte MacAusland, Louise Yeung, 以及 Lulu Zhang，他们尽心尽力为我们的报告整理大量资料与新闻故事，并耐心地进行相关编辑工作。最后，我们感谢日本国际交流基金工作人员给予我们的支持，他们总是督促我们让这个计划成为丰硕的成果，参与部份的学习之旅，甚至提供场地给我们东京的大型会议室。我们特别要感谢日本国际交流基金的Carolyn Fleisher，原英树，茶野纯一，以及佐藤敦子。最后，威尔逊国际中心与日本贸易振兴机构亚洲经济研究所的同仁为这份报告提供大量协助，但所有内文由笔者负责。此报告所发表的意见仅代表笔者的个人意见，而不代表威尔逊国际中心与日本贸易振兴机构亚洲经济研究所。

报告摘要

中国面临多重的水资源危机，包括没有妥善管制的工业废水和都市废水有毒物污染河流和湖泊，超抽地下水和表水造成严重缺水，滥伐和对湿地的破坏造成水灾。中国的水质恶化和缺水问题造成人民被迫迁徙，健康受威胁，和食物安全的问题。长远来看，水资源的问题对中国的社会，政治和经济稳定性有潜在的影响力。

面对中国水资源挑战，核心议题是如何保护河流生态系统。和缓中国水危机的当务之急让许多国内外团体投入强化法律，政策与项目的工作以提倡流域综合管理(IRBM)的观念，以及较全面防止污染的策略。执行流域综合管理的一项核心策略是中国政府对于流域管理委员会的改革。

赣江源头的渔船。赣江源头的水质仍相对干净。
照片提供：肖齐平。

这种由上而下的作法对于流域管理的改革很重要，但同等重要的还有人民与非政府组织参与决策过程，并监督流域保护发展的工作。有一些国际环境非政府组织已经在中国成立流域保护计划，结合政府，社区和非政府组织，创造多重利益相关方(multi-stakeholder)的项目以达到保护当地河流的目的。

美国和日本的政府与非政府组织各自独立在中国进行保护水资源与河流的计划，但是有许多计划往往是小规模，而且维持不久，因而限制了这些计划可以为中国真正流域综合管理(IRBM)带来制度性改变的能力。为了能更有影响力地在中国提倡流域综合管理(IRBM)，美国和日本可以在许多领域共同合作，包括集水区管理，财政和利益人相关方(stakeholder)的参与。

这份报告旨在提供美国与日本(及其它国家)的政府，非政府组织与研究机构，在中国从事流域管理合作时，所能考虑的一些选择。为建立中国水资源国际合作讨论的基础，第一部份首先探讨中国的水资源问题。第二部分回顾目前水资源相关法令与机构的效能，以及在中国日渐发展的保护水资源草根非政府组织。第三部分呈现国际援助在中国如何提倡可持续流域管理，并点出可以更加强的领域。第四部份的结论提供一些可行做法给予美国和日本政府，非政府组织和研究机构(不论是共同合作或是平行项目)在中国推动可持续流域治理。

第一部分：中国的河流危机

过去二十五年以来，中国经济发展的奇迹让上百万人民脱离贫穷，但也对环境造成了破坏。中国环境问题的数据揭示了可能发生的残酷现实。全世界二十个污染最严重的城市中，有十六个在中国。中国能源消耗量（绝大部分是低等级的煤炭）和温室气体排放量仅次于美国，高于世界任何其它国家，并且有可能在二十年内超越美国。燃烧煤炭所产生的酸雨影响着占国土面积三分之二的地区（也影响到韩国和日本）。百分之十五至二十的动植物濒临灭绝。北方严重缺水，致使人民成为"环境难民"，纷纷离开沙漠化的农田。超过 75% 流经中国城市的河流既不适合饮用，也不适合垂钓。国家环保总局的副局长潘岳指出，中国的环境恶化每年给国内生产总值的增加造成约8%的损失，也使中国的经济发展奇迹显得更加不可思议。在众多环境问题之中，严重缺水，水污染恶化以及河流环境的不当管理是对经济，环境和人民健康的主要威胁。

中国水资源的悲哀

中国政府也越来越将水资源保护及水污染控制列为工作重点。但经济发展的速度，人口增长的压力，以及地方执法能力的不足，都拖延了改善水资源问题的进度。

水资源匮乏

中国人均水供应量比世界平均水平低25%，而且预计到2030年人均供水量将从2200立方米下降到1700立方米以下，符合世界银行对于"水资源匮乏国家"的定义。水资源匮乏问题在中国北方最严重，年人均水资源占有量为750立方米。农业用水占水资源使用的80%，工业和家庭用水量也快速上升。因为这三个部门都急需用水，加上又缺乏节约措施，使得水资源更加匮乏，尤其是在干燥的北方，原本就只有中国24%的水资源，但却要生产超过中国国内生产总值45%的粮食作物。

中国北部和西部的过量取水与土壤退化造成每年平均1300平方英里土地的沙漠化，影响着4亿人的生活。中国北部和西部不断加剧的沙漠化导致农业生产无法继续维持，约2400个村落因此而被荒弃或部分荒弃。沙漠化也使得春季的沙尘暴更加严重，从2000年到2009年预估会有100场沙尘暴，远远高于上世纪23场的纪录。而且沙尘暴不只影响到中国，韩国和日本，甚至波及美国西岸。

每年对水资源匮乏与水污染给农业生产造成的损失的估计在世界银行的240亿美元到中国新闻媒体所报导的82亿美元之间。缺水问题在中国北方最为严重，但也对阻碍了全国的可持续发展。在中国640个主要城市中，有400个城市缺水，因此在2003年造成的工业损失为280亿美元。在中国的农村地区，约有6000万人在每天的日常生活当中都要面临缺水之苦。从黄河的例子可以了解水资源匮乏的严重性，从20世纪90年代中期，黄河枯竭的问题十分严重，以致每年断流达200天之久。

必须注意到，中国应对水资源短缺，增加供给的主要措施是修建水坝和引水工程。这些大型工程，尤其是水坝，所费不赀。由于可能因此失去土地和生计，此类水利工程受到越来越多地方居民的反对。一些研究估计显示，更严格的水资源保护手段可以帮中国每年节省1-2千亿立方米的水，从而使现有用水量降低1/4，并因此避免了修建一系列昂贵且备受争议的水利工程的需要。

水污染

中国所有的主要河流都深受水污染之苦，影响人体健康及工业生产，也破坏河流的生态系统。工业调节的松散和水资源协调管理的缺乏是导致中国严重水污染的两个体制失误。从2002年起，每年约有630亿吨的废水流入中国的河流，其中62%是工业污水，38%是稍经或未经处理的城市废水。

废水处理是第十个五年计划（2001-2005年）的重要项目，然而根据中国国家环境保护总局的监测，2001年以来建立的污水处理设施只有一半真正发挥功用，而另外一半则因当地政府觉得设施维修费用太昂贵而被迫停止运作。到了2002年底，只有39.9%的城市废水经过处理，在农村废水处理的比例更低。全中国可能只有约20%的废水经过处理。

反应中国水资源保护工作不足的一个最明显的例子可说是淮河。尽管从1993年中央政府便已经开始了长达十年的整治工作，但淮河仍是中国水污染最严重的河流之一。严重污染的淮河直接导致淮河盆地人民的高致癌率和对于健康的危害。举例来说，住在淮河盆地一些村庄的青年，许多年来都因为健康水平低而无法通过征兵体检。

恶化的水体生态系统

水污染和过度取水是造成中国河流系统恶化的两个主因，但是许多河流，包括长江，还面临很多其它问题，比如过度砍伐森林，将湿地变更为农田，不当建设开发冲积平原，这些因素都造成更严重的水患。此外规划不合理的长江水利工程扰乱了河流原本的自然流动，破坏了流域盆地的生态系统，并造成生物多样性和河流生产力的损失。

除了水资源质量退化所造成的环境和经济损失外，水污染和缺水问题也造成了中国社会的动荡。西方媒体与非政府组织常常把焦点放在一些知名的水资源冲突案例，比如长江三峡沿岸居民拆迁的问题。但其实跨省和地方性冲突的数目一直在增加，而且有越演越烈的趋势。

流域综合管理的三个核心要素

藉由检视中国河流所面临的问题，我们可以了解中国日益恶化的水资源危机，也可以了解对于实施有效流域管理造成阻碍的政治与社会问题。尽管几乎全世界的国家都面临保护水资源的多重挑战，但是中国的问题更为严峻，因为中国目前未能可持续地管理水资源，尤其是河流。失控的开发和水资源管理体系协调性的不足产生了很多不利影响，因此强调了中国在河流管理中采用全盘方法的必要性，也就是所谓的"流域综合管理"（Integrated River Basin Management, IRBM）。

在流域综合管理复杂的观念中，我们相信有三个关键的机构是中国政策制定者，非政府组织与国际援助提供者在提倡流域综合管理观念时，所必须注重加强的，这包括，(1)流域管理机制，(2)财政机制，和 (3)公众参与的机制。以下是这三个机构在中国的简短回顾。

位于赣江中游的一家造纸厂没有采用合适的污水处理措施，而是建造围栏将布满泡沫的污水收集起来，进行稀释，然后排入赣江。照片提供：肖齐平

分散破碎的管理机构

中国水资源的核心问题是水资源管理协调性的无效和不足。1988年中国第一次通过水资源保护的法律及相关条例，内容涉及水费征收，水资源分配，水资源使用许可和节水设备安装。然而仍有许多问题阻碍水资源管理的改革，包括地方政府和流域委员会监督和落实能力的不足。2003年水资源修正法中的一个核心目标是强化流域委员会，让他们可以有效施行水资源保护和管理的议题。

目前中国的七个水利委员会 (river basin commissions) 成立于50年代，目的是为了开发水资源，发电，缓解水患，及提供航行设施。作为水利部的分支机构，水利委员会有很强的技术与水利专门技术，但往往缺乏监督和落实水资源保护的能力。大体来说，中国的水利委员会对河流的生态健康不够注意，对上下游水资源需求的平衡也未能充分协调。

水利委员会的另一个局限是它与其它政府机关，省政府和地方政府之间缺乏良好的关系 (有时候甚至是敌对的关系)。另外一个水利委员会的问题是他们无法在流域管理的行动中联合广大的利益相关方。

有限的财政工具

通过流域管理的计划与建设分配水资源时，水利委员会和其它政府机关往往没有充分考虑经济、社会和环境的成本。虽然中央政府在倡议活动和五年计划中制定了一些宽松的清洁河流与净水目标，但是提供给地方政府的经费往往不足。举例来说，十年的淮河清洁计划并没有实质效应。虽然第十个五年计划有许多治理水污染的目标，但是政府为环境保护提供的经费却短缺30%。

中国与其依赖中央政府的补助，不如提供经济刺激，促使产业与地方政府保护河流生态，尤其是下游的使用者应该补偿上游对河流保护付出的努

力。中国的流域委员会与城市缺乏经济机制来为迫切需要的废水处理设备提供财源，例如循环基金和债券。一个主要障碍是水费的低费率以及水费征收的层次太低。许多市场工具，如绿色税，水权交易，或是上下游的互相补偿机制都可以推进河流资源保护，但在中国的发展仍然很缓慢。

水资源政策的制定缺乏透明度和公众参与

不论中央或地方，在设计发展水资源管理的计划时，政府机关都不通知民众或询问民众意见。水资源政策的公众参与局限在民众的抱怨和抗议。因环境污染事件受害的中国人民可以向政府提出正式申诉，控告污染制造者(甚至是政府机关本身)，但这些努力所能造成的改变仍有限。为增进中国河流保护，将需要更多民众与非政府组织的参与，来监督水资源的政策与措施的实施。

美日在中国成为水资源伙伴的潜力

中国的经济改革加速了工业化进程，提高了生活水平，并将广大农村劳动力从农业生产中解放出来。但这些发展也造成了环境的恶化，并直接影响着中国人民的健康及中国的经济发展。中国经济的全球一体化虽然加速了经济增长，但也导致了环境恶化。在这样的前提下，美日与中国在各个领域的环境合作显得更重要。虽然中国水资源的挑战很艰巨，但也为美日在集水区管理，财政和利益相关方的参与合作方面提供了大量的机会。

第二部分　中国国内江河流域保护的努力

日渐严重的缺水和污染问题促使中国政府更新旧法令或是出台新法令以解决水资源管理，污染控制及流域委员会体系改革等问题。保护水资源也在最近的两个五年计划中成为投资重点，因为可持续的供水是经济增长的关键。这些由上而下的做法对于水资源管理法令与机构的改革很重要，并将完善水污染防治的硬件设施。同样重要的是政府要不断扩大政治空间，让人民与非政府组织可以对水资源部门工作由下而上的参与。以下我们首先简要叙述中国政府在水资源管理与保护工作上所作的投资，法律体系和机构设置。紧接着的是我们会指出一些公共参与的窗口，让公众，特别是非政府组织得以参与水资源保护活动。

以从上而下的作法，减轻中国水困难

水资源在政府高层被放在优先地位

自2001-2005年的第十个五年计划中，政府制定了令人称羡的环境保护目标，尤其是围绕水资源保护。但是在承诺的8500万美元投资中，30%资金最终未能到位。在第十个五年计划中，政府计划要在淮河，海河，辽河流域和太湖，巢湖和滇池流域建立或扩建145个城市污水处理厂。除此之外，城市废水处理率应提高到50%。虽然中国政府在工业废水处理上有所进步，但环境保护总局对污水处理厂的检验显示在第十个五年计划间建立的污水处理厂中，只有一半在运作。

在第十个五年计划期间，政府继续执行在中国南方禁止滥砍滥伐的法令，并对坡地的退耕还林进行投资。这些项目缓和了加剧1998年长江特大洪水的土壤侵蚀的情况。2005年底，环境保护总局宣布水污染控制的努力已使水质得到提高。但此论断和可能高估了所取得的成果。举例来说，在清理淮河的十年计划中，环境保护总局声称水质大幅改善，但项目最终不得不承认淮河污染仍十分严重。中国一位高级的水资源专家也曾指出第十个五年计划要在2005年底达到清理淮河的目标是不可能的。

2005年10月初中国共产党提出第十一个五年计划(2006-2010年)的拟案，并将在2006年　3月在人民代表大会中通过定稿。目前的草案概括了保护自然生态系统及能源的纲领。计划其中的一部分明确地提出应该要制定保护湿地，修复中国沿海严重受损生态系统的环境政策。这项五年计划也包括以劝导机制来加强三个主要河流(淮河，海河，辽河)和三个主要湖泊(太湖，巢湖，滇池)，以及长江三峡大坝区域，长江与黄河的上游地区的污染控制工作。这项计划也将南水北调工程沿途的污染控制视为重心。国家环境保护总局计划将所有城市废水处理的比率从45%提升到60%。前环境保护局局长谢振华指出第十一个五年计划环境保护的总投资金额将达到1566亿美元(1兆3000亿人民币)，超过国内生产总值的1.5%(GDP)。

中国的水质管理

虽然中国政府形成了颇完整的一系列水资源保护法令,但主要的体制问题仍然阻碍了水资源管理的有效实施。这在河流管理方面尤其明显。由于水资源归国家所有,被视为公开共享的资源,因此水质管理的一个主要挑战是缺乏明确的水权制度,清楚界定的水权划分的缺乏制约了水资源保护工作以及水权的交易。而水权交易恰恰可以让水资源在不同部门之间交换而不发生冲突(其中是将水资源从农业分配到工业)。

另外一个中国水资源管理的体制性挑战是中国国内存在管理水资源的不同政府部门,亦称为"多头龙"。主要的竞争者有负责水资源品质的水利部和负责监督水污染控制的中国环境保护总局,农业部,建设部,与渔业部也涉及部分的水资源管理。但整体而言,水利部在处理中国的河流流域管理以及水资源品质控管上享有最高决策权。值得注意的是,尽管中国过去二十五年一直在分散中央集权,但在水资源管理上却一直重新加强中央集权,将其纳入水资源部及七个主要水利委员会的辖下。

中国在1988年通过了第一个《水法》,目的是明晰水资源管理的权利,并创造让水资源保护机构能够更清楚地定义水权范围的结构。举例来说,这部法令包括,(1)提高水费的规定;(2)一套用水许可制度,以定义工业,农业和城市用水许可,因此可以让地方政府征收更多的水费;(3)流量分配,以便在面临缺水问题的不同省份之间分配水权。

地方政府是水费征收与用水许可改革的另一个阻碍,因为他们往往担心限制用水会限制地方经济发展。然而近年来,有一些主要城市已经开始提高水费,设置更多水表。这些都是减缓地下水和地表水过量抽取的必要措施。尽管如此,缺水的城市通常仍会选择开发新的水资源,而非严格执行水费征收,用水许可或其它节水方式。在面临缺水问题时,北京市政府也是从供应面解决问题。

一个在地方提高水资源管理的有效方式是建立城市水利局。这个新的单位可以结合当地的环境,水资源和城建单位,共同来管理水资源,并解决多头龙的问题。另外一个还未列入法律的可能性,是在县之间,或是工业与农业之间,进行水权交易。浙江省在2000年进行了第一次水权交易,由义乌市(县)向水库上游的东阳市(县)购买每年5000万立方米的水。

90年代中期,由于来自水利部与地方水利局的压力及努力,许多区域开始颁发用水许可,但有限的数量并没有促成迫切需要的水资源保护,而且也常常没有被确实执行。在几个流域所开展的水资源分配计划也没有完全成功,主要原因包括,(1)水利部和省政府没有设立执行机构;(2)流域组织缺乏能力和影响力来协调与监督水资源分配的工作。

2002年修改的水法旨在填补过去法令的漏洞,并要求水利委员会用总量控制的策略(从整体全面的角度来分配流域内所有水资源,并保留一些水资源作为生态流),在河流流经的省份间分配水资源,以解决日渐严重的缺水问题。切实执行水资源分配制度意谓着地方政府必须有效监督并执行用水许可,把水资源保护放在工作的首要地位。但是要管理流域的水资源,不论是程序上或政治上,都是很不容易的,也因此在所有七个主要水利委员会当中,只有最强大的黄河水利委员会成功建立实施了一个颇为成功的水资源分配制度。

管理水资源品质的挑战

1984年中国政府通过《水污染防治法》,并于1996年对该法作出修改。这项法律规定在各级政府的环境保护局都应该进行水污染防治的管理与监测。但尽管存在对污染进行约束的法律,但地方政府对于工业采取的保护

主义使地方环境保护局无法有效执行水污染防治法。举例来说，尽管污染收费很低，但地方政府仍经常将高达80%的收费退还给排污工业。由于工业企业是地方政府财政税收的重要来源，地方政府对于它们的非法排放物也常常睁一只眼，闭一只眼。

《水污染防治法》和《水法》中相互矛盾的条款加剧了各政府部门间的冲突，也让水资源品质计划、保护和监测的有效执行工作更困难。尽管环境保护局通过1989年的《环境保护控制法》和《水法》来管理水资源的品质，但《水法》又规定水利部是管理中国水资源的主要单位，因此水利部将保护水资源质量视为其责任。这样，水利部和环保局就水污染管理存在激烈的政治斗争。

环境保护总局和水利部之间的冲突造成的一个主要问题就是两个部门会为水资源质量的数据搜集和水污染管理的权限竞争，造成重复浪费，也让清理水资源的工作严重缺乏合作。与水质问题相同的是，地方政府缺乏彻底执行水污染相关法律的意愿，而中央政府也缺乏督促他们的能力，这些问题都造成了中国河流流域严重的环境恶化。

全国人民代表大会也在2006年初完成《水污染防治法》的修定，增加加强执行能力，明晰水污染防治责任权限的条款。

实施流域综合管理的障碍

尽管2002年的改革在书面上大大地加强了水利委员会的权利，这些机构仍需要更多的改革及能力建设，才能具备实行流域综合管理的完全能力。目前中国的流域委员会都仍只是水利部的分支，采取由上而下，且较狭隘的流域管理方式。除了缺乏完整的水资源质量管理机构以外，中国的流域委员会会也缺乏让地方政府与人民参与的机制。事实上，"水利委员会"的名称会让人产生误解。中国的流域委员会并没有任何委员或正式机制让省级和地方政府在由上而下的流域管理机制中参与政策决策和权力分配。。

因为缺乏正式参与的角色，使得许多省政府只好绕道走后门进行协商，而这对有效流域管理进一步造成障碍。举例来说，2002年黄河几乎断流，使得最下游的山东省面临严重缺水，影响秋收。山东省政府派了代表到北京，成功游说了从内蒙古的蓄水池引水。但这却造成上游的严重缺水。这种因状况而异解决水问题的决策方式只会造成更多的冲突，而没有真正促进水资源与生态的保护。

新环境影响评价法

过去几年来，环境保护总局以扩大公众参与来作为保护自然资源与人民健康的另一渠道。1979年中国的第一部环境保护法模糊地叙述了人民影响环境政策形成与执行的权利。然而直到2003年通过的《中国环境影响评价法》，才明晰并强化了中国人民影响环境法律与基础设施项目建设的权利。90年代中期中国通过第一部《中国环境影响评价法草案》，但是只适用于土建项目。新的法律要求针对基础设施及其它各种建设计划进行评估。而且环境评估法报告也必须向大众公开并允许民众提出意见。

新的《环境影响评价法》让环境保护总局可以实际发挥其环境保护的职能。举例来说，2005年1月环境保护总局推迟了全国30个大型建设计划，其中有许多是水库与相关水利基础设施项目。推迟的原因是因为这些项目没有依法准备适当的环境影响评价报告。大部分被推迟的计划规模都很大，包括长江上游金沙江沿岸的溪洛渡水力发电厂。

直言不讳的中国环境保护总局副局长潘岳则表示，即使环境保护总局可以初次"胜利"推迟这些建设计划的执行，也不代表环境保护总局具备完全的能力去检视所有的建设计划。潘岳强调说民众参与环境评估的过程也非常重要。他指出环境保护总局计划要举办听证会和论坛，以方便公众参与。因为全国环境保护总局和各地方环保局仍缺乏明确的程序来举办听证会及论坛，有一些国际非政府组织，如美国律师协会就如何举办听证会和其它公众参与机制提供了培训。日本国际协力机构也协助环境保护总局起草环境影响评估中公众参与的实施纲要。中国政府不只很欢迎国际组织对环境保护法规与项目的协助，也逐渐给予国内非政府组织更多空间推行环境保护及教育的工作。

由下而上的行动以面对中国水资源的困难

不足的一环—水资源保护和流域管理的公众参与

2005年11月13日，中国石油天然气股份有限公司在吉林省的化工厂发生爆炸，将上百吨的苯排入松花江。松花江流入黑龙江省，是省会城市哈尔滨的饮用水来源，也是下游600公里俄罗斯哈尔罗夫斯克市饮用水的来源。开始几天吉林省与地方政府都没有向下游的政府或全国环境保护总局报告。一旦报告了以后，哈尔滨的官员又设法掩盖事实，在事件发生十天以后通知民众供水将因为"正常维修"而中断。但是随着城市中关于化学物泄漏的流言越来越多，市政府官员迅速改口说供水将中断四天以避免民众接触苯。由于对地方政府官员提供关于苯造成的健康危害的正确信息的怀疑，许多人离开了哈尔滨。

中国新闻媒体一开始快速且强烈地批评地方对于此危机的反应失当，但几天之后负面报道减少，转而正面报导中央政府的努力(包括其对事件的调查与对地方官员纪律的严肃)。全国环境保护总局的局长解振华被要求辞职。这个案例显示了工业企业享有的地方保护主义，危机处理的缺陷，政府透明度的不足，以及通知及团结公众参与环境保护机制的缺乏。

环境非政府组织的登场

从1994年开始，中国环境政策开始扩大公共参与，因为新的行政法规允许"社会团体"的注册登记。中央领导人允许文明社会有较大的政治空间，因为他们意识到政府需要更多人民的帮助才能解决经济快速发展与社会福利制度瓦解所带来的社会与环境问题。但是无可否认的是，政治参与空间仍有一定限制。注册登记的限制在于要求非政府组织在申请登记时，必须有一个政府单位的督导(被称为"婆婆")，而且不可设分支办公室。另外一项限制非政府组织的法律障碍是在同一个城市或省份，不可有两个以上的组织致力于同一目标或进同样的工作。

1994年第一个在新法令下登记成立的是环境草根组织—自然之友其它的环境组织也开始申请成立，而那些无法成功注册的则登记为商业团体或是干脆在没有正式注册的情形下运作。越来越多的环境非政府组织只登记为网络团体，因而能免除所有登记注册的程序。至今，中国有将近2000个环境非政府组织，且成为中国文明社会发展的前锋。一开始，中国环境非政府组织趋向于倡导一些较"安全"的议题，例如学校的环境教育或是向大众宣导资源回收，水资源保护和动物保护等题目。

尽管有注册登记的挑战和避免冲突的压力，到了90年代后期有一些团体

在中国南部各省，很多河流和湖泊由于生活污水和农业面源污染的大量排放而出现了严重的富营养化，致使一些有害外来入侵物种泛滥成灾。图为中国东南部福建省一个长满水葫芦的湖泊。（照片提供：邓佳）

开始进行地理范围与议题的扩展，并进而提高他们对于政策的影响力。中国大部分的环境非政府组织都位于城市，或是在四川省和云南省进行生物多样性热点的保护，但也有一些组织有效地进行保护水资源的工作，尤其是河流流域保护以及公共参与。

非政府组织参与水资源工作

虽然为数不多，但一些中国的非政府组织已经开始从事河流流域保护的工作，且很多都将公众参与作为工作的重要部分(请参见表三)。一个独特的非政府组织是污染受害者法律帮助中心(CLAPV)，在协助污染受害者的工作方面扮演重要角色。虽然1979年的环境保护法(1989年修定)让污染受害者有权利向污染者提出起诉，但实际上要由民众自己进行这个法律程序确实很大的挑战，尤其是在地方政府特别保护工业企业的情况下。在过去几年中，私人律师一直在帮助重大水污染时间的受害者赢得诉讼。参与污染受害者法律帮助中心的律师们正在建立先例，敦促法庭提高能力建设，以处理此类通常要求法官具有专业水平的案件。

中国环境非政府组织发展的一个重大发展分水岭是2004年与2005年，环境保护运动人士与记者共同建立了一个全国促进怒江13个水力发电水库建设信息更透明化的行动。2004年秋天，中国的一些环境运动人士得知云南省政府要在怒江建坝，他们于是安排一群记者前往流域采访，了解水坝的建设计划，以及对于周围环境的影响。

一旦第一批记者开始报导怒江美丽的景致，及其世界遗产地位，更多的记者开始蜂拥而至。数周内，全中国有数以百计的新闻报导与广播谴责水库建设以及水库计划的缺乏透明，尤其是这项建设并没有进行环境影响评估。环境保护运动人士建立了"中国河网，"以协调大家的工作，在全国进行摄影作品巡展让大家可以看见陷入危机的怒江是多么美丽，并向给中央领导人寄出请愿书。

这场广泛的大众辩论引起中国总理温家宝的注意，并在2005年2月下令在环境影响评估报告完成之前，暂停水库的计划。2005年8月总共有61个非政府组织与99个研究机构与政府单位共同组成一个联盟，向最高领导人呈递了一封公开信，要求政府在批准建设之前，必须公开怒江水力发电水库计划的环境影响评估报告。

37

报告完成之后，围绕水库的辩论仍持续在进行。即使水库的建设最终动工，这场环境运动仍代表中国环境保护人士的巨大胜利，因为他们与记者合作，将这项议题变成公开的辩论。这场环境运动是十年来中国非政府组织与中国政府一起合作的成果(且往往不是对抗)，也证明了中国环保人士不断扩大的自由空间。与其说这些环境非政府组织在怒江的行动是一场"反水库"的行动，还不如说是中国水资源管理与环境中政策制定更透明，多公共参与更广泛的一场运动。

结合由上至下与有由而至的有效策略

大部分的国家都面临执行水资源保护法的挑战，而中国所面临的挑战更加艰巨，包括人口增长压力，经济快速增长，水权定义模糊，以及地方政府的保护主义。中国政府建立了强大的法规来防止水污染并强化水资源保护，政府也与国际多边组织及国际非政府组织合作，共同解决水资源管理与河流保护的问题。

除了邀请国际专家以外，中国的领导人也渐渐允许中国环境非政府组织有更大的政治空间，因为他们了解政府无法由上至下地解决所有的环境问题，尤其是水资源的问题。过去几年来，环境保护总局的官员一再强调公众必须参与环境法规的决策与对地方政府与企业的监督，因为只有通过这种由下至上的参与才能减轻政府执行环境保护法规的规划与财政负担。2004年7月1日国务院通过《执行许可法》，要求行政当局必须让人民知道他们在听证会上有发表意见的权利，可以针对任何可能影响到他们的政府计划提出意见。环境保护总局是中国第一个将举办听证会纳入法律的政府单位。

另外一个环境领域透明度提高的标志是2005年秋天环境保护总局向各方征求关于如何提高环境影响评估过程中的公众参与的建议。这项新的法规包括保护参与者权利，信息公开，并设计公共参与的机制等条款。这是中国政府第一次公开地寻求收集大众意见，并将其列入制定新法的考虑。

2005年11月环境保护总局呼吁全国实施企业的环境表现评比与信息公开，这也是环境信息透明化重要的一步。2005年12月国务院在《环境保护的决策》一书中，包括要求企业必须向大众公开环境信息的条款。

尽管有一股洪流推动着环境信息的更公开及透明化，但环境领域中非政府组织与民众参与的发展，也让政府对中国社会越来越多的社会运动产生顾虑。全中国内容广泛的大量抗议行动让中国政府开始担心。地方政府对于当地非政府组织监督污染工厂的行为格外小心。不过还是有一些市政府，包括深圳，北京和厦门，都非常欢迎非政府组织与市民参与环境保护。也有许多国际计划倡导公众参与如何可为政府达成环境保护的目标，并减少社会纠纷。只是中国的环境非政府组织不但面临许多外部局限，也有许多内部挑战，而这些都有可能会伤害非政府组织长期的可持续性，包括对于国际援助的过度依赖，缺乏内部透明度以及因为低薪资而造成人员流失。

为了让中国政府可以强化其水污染与水资源管理的法令，中国政府不只应该持续由上至下改革法律及水利委员会，也应该推动改革来促进环境非政府组织的发展及公众参与。一些必要的改革包括，(1)修改法令让非政府组织的注册更容易;(2)推动免税捐赠的相关法令，以鼓励中国企业与人民向当地非政府组织捐款，并藉此让中国的非政府组织可以脱离对国际援助的依赖;(3)让公众与非政府组织可以更容易获得环境决策(像是新的环境评估法)与计划执行的相关信息。

第三部分　保护中国河流的国际援助

在过去二十年来,许多国际组织与国务院,全国人民代表大会,环境保护总局,水利部以及其它部门合作,出台了新的环境政策,管理法规和试点项目。

然而从二十世纪90年代开始,水污染所造成的健康危机及冲突促使中国政府开始向国际社会寻求援助。以下我们提供了国际社会在中国所进行的水资源和河流保护项目的概况。尽管在此领域的国际活动不断增加,在中国仍存在很多开展类似活动的机会。第四部分对此将有详细叙述。

国际社会在中国的流域保护相关项目

多边援助

世界银行

中国是世界银行在环境领域中最主要的援助金与贷款接受国,受援助项目包括空气污染控制,草原保护,环境数据公开,以及水资源保护。世界银行参与了多个水资源保护项目。其中两个备受关注的项目直接以提高塔里木河与海河流域管理机构的能力为目标。

塔里木河流域。世界银行在新疆进行了一项极具挑战性的项目,就是在塔里木河流域创造一个新的流域管理委员会。这是中国第一个完全 "参与式" 的流域管理委员会。虽然要将这个项目的经验运用于面积更大,涉及利益相关方更多的流域十分困难,但是对其他国际组织在中国政府管辖范围内,在较面积较小的流域开展支持类似体系的项目却很有帮助。

海河流域。2004年世界银行通过其全球环境机构,提供1700万美元的援助金,在海河流域实施旨在加速流域水资源和环境管理一体化的项目。该项目面临的最大挑战是迫使国家环保总局和水利部共同实行体制改革,以建立地方水利部门和环保部门的合作机制。该项目还着眼于提高水资源综合规划所需的技术支持。

亚洲开发银行

从1986年以来,中国成为亚洲开发银行的第二大会员,而且具有最好的金融资产表现。亚洲开发银行所援助的环境项目范围很广泛,包括提高能源效率 (可更新能源),城市环境保护,以及水资源管理的改革。很多水资源的相关项目将重心放在城市水资源管理 (包括废水处理与供水系统),湿地保护 (三江平原)以及流域污染控制 (海河)上。从2003年开始亚洲开发银行在黄河流域开展了一项主要研究,名为 "跨管辖领域环境管理项目"。这个跨部门的研究将焦点锁定于黄河保护的法律,财政,管理与社会问题的层面。项目第一阶段分别从国家及地方级别调查水资源管理的法律及实践。该研究还特别围绕跨管辖领域水污染纠纷话题,对政府部门间的关系协调机制作出了评估。第一阶段的研究建议中国政府修改水资源相关法律,并建立新的立法与合作机制 (如跨部门委员会),以完善流域内机构之间的合作及监督

表1 活跃于中国水资源工作中的美国非政府组织和大学

• 世界自然基金会-中国（WWF-China）在长江流域有多个流域综合管理活动项目。其展示项目包括如何通过恢复湿地和湖泊，如何通过社区教育和非政府组织能力建设的活动扩大公众对水资源管理的参与，提高洪水控制能力。2005年世界自然基金会成立了一个小型拨款项目，以资助22个旨在保护长江水生物种的项目。

• 从2005年开始，国际保护（Conservation International）与自然保护和中国国家林业局和作，在中国西南发展碳与水资源的环境服务赔偿体系。其中一个正在策划试点项目设在云南丽江（为国际保护森林与气候社区以及生物多样性项目的一部分）。项目将联合上游的农民和下游的用水方共同开展流域保护和在造林的工作。国际保护还与全国人民代表大会的环境及自然资源保护协会合作进行研究与项目开发，帮助宣传中国已产生的环境服务赔偿立法。

• 根与芽（Roots & Shoots）是简·古道尔（Jane Goodall）学会中国分会的一个项目。学会的中国分会将在2006年初与成都城市河流协会（一个中国非政府组织）合作，在四川农村地区开发一个"生态村示范项目"，作为流域清理计划的一个组成部分。在清理城市供水源的努力中，成都城市河流协会在流域上游积极活动，解决农业生产中化肥和杀虫剂流失的问题。成都城市河流协会，四川大学，简·古道尔学会中国分会及根与芽将与上游一个村庄合作，通过同时开展环境教育，生态农业以及当地生计，以综合管理的方式解决农业化学物质流失的问题。

• 自然保护（The Nature Conservancy）与中国政府和相关学术机构合作，正为有关中国生物多样性分布，代表和生存能力设计一幅综合的科学地图。在该活动中，国家发改委和国家环保局首次合作开展同一计划，以知会可持续经济发展的政策决策，并重新设计和扩展中国的保护区系统。作为合作项目的一部分，自然保护将开发全面的数据系统，评估和监测约有3.5亿人口的长江上游地区的淡水生物多样性，并在该地区建立资源保护的重点和策略。自然保护还帮助促成了一个综合电网内可持续能源构成方案的评估。按照该电网的设计，水电的开发将最大程度的被限制在淡水资源维护和保持当地生计的范围内。

污水处理的比例很低（全国的处理率为４０％）。城市为流入河流的污染物的主要来源。照片提供：肖奇平。

机制。项目的第二阶段包括分析如何完善黄河支流渭河环境保护的财政机制。这项财政分析旨在为中国国务院水资源管制项目的管理及融资提出具体的建议。许多现行的水资源管制的法令都太过宽松，所以亚洲开发银行希望通过这项研究可以为水利部，环境保护总局及其它中央政府部门提供详细的技术背景，以便它们修改现行法律，健全中国河流保护体系。

双边援助

英国国际发展部

英国国际发展部的重心是通过与发展中国家的合作来扶贫。针对中国水资源面临的挑战，英国国际发展部通过完善水资源管理和提供可持续清洁饮用水，来提高人民的生活和健康水平。英国国际发展部还与其它国际组织合作，支持中国政府执行2003年水法修正后的水利部门改革，包括使用者参与的扩大，水资源综合管理的办法，新的水土保护办法，以及饮用水和卫生系统更广泛的供应。

欧盟

从2002年至2006年，欧盟与中国的合作案的预算是2亿5千万欧元，其中有百分之三十投入于环境保护与可持续发展。最大的环境项目之一是辽宁省辽河的河流保护。对重工业部门管理的欠缺以及农业排水，使得辽河成为全中国污染最严重的一条河流。向辽河的过度取水造成辽宁省严重缺水的局面。与全国人均2292立方米的用水量相比，辽宁省的人均用水量只有603立方米。五年以来，欧盟在北京的办公室在欧盟总部，日本和世界银行的支持下，与辽宁省政府合作，建立一系列促进可持续江河流域治理的项

目。欧盟与其中国的合作伙伴目前正通过以下措施，来建立一套污染控制与水资源管理的综合架构: (1) 在其中一个主要水库（大伙房水库）集水区建立试点水质保护计划; (2) 调查工业用水的保护与污染; (3) 用地理信息系统和其它决策分析的软件开发整个流域的水质模型。这些项目活动使欧盟团队能够就流域范围水利体系与水价的改革提出了建议，并在辽宁省第十个五年计划中被采纳。

瑞典国际发展合作局

多年以来，瑞典国际发展合作局所参与的水资源相关项目包括, (1) 工业节水技术; (2) 准备了对内蒙古一个湖泊恢复的综合行动计划; (3) 农业节水; (4) 完善污水处理场管理体系的能力建设; (5) 发展兼顾生态的卫生系统。瑞典国际发展合作局是中国环境与发展国际合作委员会的主要资助者，而且在委员会的工作中，水资源管理是一个重要领域。瑞典国际发展合作局也为中国污水处理厂的建设提供贷款优惠政策。从2006-2010年瑞典国际发展合作局的新合作策略也会继续将可持续环境发展放在首要位置。

瑞士的双边援助

2005年大部分瑞士的环境双边援助都集中在四川省的项目与研究。瑞士研究者与中方的合作伙伴共同进行了生态旅游项目的研究，培训和示范，活动包括四川省境内世界遗产与香格里拉山脉生态旅游的开发，以及龙泉湖与三叉湖的农业生态旅游。2006年瑞士在四川的工作包括对闽江和沱江流域水资源与环境管理的研究。研究将调查在流域预防工业，城市及农业污染的管理及政策方案。另一项与水资源相关的研究将在2007年开始，考察长江上游的生态保护建设。这些研究将探讨上游可持续生态林业对森林生态系统的恢复，以及预防长江地区水土流失和严重洪水的作用。

中国环境与发展国际合作委员会

中国环境与发展国际合作委员会是一个高级顾问委员会，由中外专家与政府官员组成，为国务院关于环境与发展的策略提出建议。2003 年3月委员会成立了一个流域综合管理 (IRBM)工作小组,工作重心是黄河。总目标包括以广泛的非政府组织参与为前提，通过完善水资源管理、保护生物多样性、以信息共享的方法管理生态系统,示范 和公众参与，以推动中国的流域健康。除了了解流域综合管理是如何在全世界推行以外,工作小组还与中国世界自然基金会(WWF-China)合作，进行对长江的研究。该保护计划的想法已在中央及地方政府部门以及社会团体中得以交流，以寻求向中国环境与发展合作委员会呈递的反馈意见,作为未来立法与试点项目的参考。

日本政府在中国的水资源相关工作

从20世纪90年代中期以来，日本对于中国的官方发展援助就集中在环境项目上。相当数目的项目都与水资源有关，尤其是污水处理，供水设施建设，大规模灌溉区域的水资源保护及流域改善。2004年日本政府宣布将通过造林，防治沙化 以及集水区管理，把水资源管理与保护视为首要重点。此外，日本还将巩固对中国长期的双边援助，对付水污染与生态保护的问题。日本政府对中国环保的官方发展援助绝大多数是对基础设施建设的日元贷款。但是，目前日本政府正与中国政府协商，考虑将从2008年北京举

表2 活跃于中国水资源工作中的日本研究中心和非政府组织

• 日本拉姆萨中心(Ramsar Center Japan)一直活跃于日本,中国及其它亚洲国家有关湿地的研究和公共认识工作。在中国,日本拉姆萨中心与设在国际湿地联盟的背景办公室合作,为中国,韩国,日本的中校学生提供有关湿地保护的环境教育和交流项目(2004年在江苏大丰举办,2005年在黑龙江扎龙自然保护区)。

• 2005年日本环境会议的一些成员访问了河南省,以了解与当地一个独特的非政府组织淮河卫士(第二部分方框2又详细介绍)在淮河水质污染方面的合作机会。

• 湄公瞭望(Mekong Watch)是位于东京的致力于湄公河守护和政策研究的非政府组织。2005年湄公瞭望派遣了一支工作组至云南昆明,与当地名为绿色流域的非政府组织合作,共同研究建坝和湄公河上游的其它开发活动对云南省湄公河流域及下游可能造成的环境威胁和对人民生活的影响。

赣江景色。
照片提供:肖齐平。

• 日中新世纪协力与其中国合作者中华全国青年联合会合作,在北京(2004年4月)和札幌（2005年10月）举行了日中水资源论坛,目的在于增进两国政府官员,学者,工商业代表及非政府组织积极分子围绕水资源保护话题的交流。

• 从2004年开始,日本水论坛开始举办日中水问题学习会,邀请日本和中国的水资源专家参加会议,交换信息,以促进两国间关于水资源问题的相互了解。

在长江第一弯（金沙江段）开始建坝的行动引起了广泛的关注。建设中的水坝并不是位于长江支流，而是位于长江的主要干流。在整个金沙江段计划建坝12个，工程将终止于虎跳峡。图中的水坝为这一系列水坝中的第一个。计划建设这些水坝的目的除了水力发电外，还有防止三峡水库过渡淤积。(照片提供：马军)

办奥运会的年度开始，终止日本对中国的日元贷款援助。如果日元贷款援助终止，日本对于中国的官方援助将会较着重于制度性改革及人力资源开发，而不再是基础设施建设。

日本国际协力银行

日本国际协力银行以日元贷款形式援助发展中国家。在中国有三个主要重点：环境，人力资源开发 以及西部的扶贫工作。从1979年开始，日本国际协力银行 （前称为海外经济协力基金)开始向中国提供大量贷款。过去五年以来，日本国际协力银行所提供的贷款每年平均为150亿美元。日本国际协力银行虽然没有特别针对流域管理的项目，但却参与了许多水资源相关项目，例如 (1) 中国20多个大型城市的供水项目; (2) 其参与的水污染控制项目支持工业污水处理和排污厂的建设，及其在河南省的淮河流域，吉林省的松花江-辽河流域，黑龙江省的松花江流域，湖南省的湘江流域 和重庆市的三峡大坝上游的扩建; (3)新疆和甘肃省的节水灌溉设施; (4) 黄土高原的造林(陕西，山西和内蒙古)以缓和黄河淤积的情况; (5) 长江中游湖北省与江西省的造林; (6) 四川省，河南省与其它省分的防洪与供水多功能水库。

日本国际协力机构

日本国际协力机构的水资源相关项目包括技术合作项目，派遣日本专家对中国同行进行训练，或是邀请中国专家赴日接受训练。训练领域包括, (1)水资源项目的人力资源开发。日本国际协力机构的目标是对2000多名中央和地方政府的水利部门人员提供培训; (2) 大型灌溉计划节水措施的模式计划项目; (3) 太湖的水环境重建试点计划; (4) 四川省的示范造林计划。日本

国际协力机构目前正在新疆进行土鲁番盆地可持续地下水的开发研究，还在云南省的小江流域(长江上游的支流)进行山体滑坡灾难控制的综合研究。此外，日本国际协力机构还与中国建设部，水利部和其它地方及省级政府合作，拟写一份普及节水灌溉的手则。

尽管日本大部分的水资源相关援助都专注于技术转让（水资源管理硬件），但过去几年开始，日本开始出现了一些着眼于人力资源与水资源政策(需求更甚的水资源管理软件)的项目。此外，日本国际协力机构刚刚在中国启动了一个水权项目，由日本国土交通部与相关大学的专家学者提供协助，针对辽宁省的太子湖进行个案研究。作为国际合作机构倡导的大型灌溉计划节水措施模范计划项目的一部分，日本的水土综合研究所与中国进行技术信息的交换，目的是分享日本包括土地改良区的经验。土地改良区在日本已经施行超过四十年，被视为一项成功的参与式灌溉管理经营模式，提供了许多值得中国学习的经验。

美国政府在中国的水资源相关工作

和日本全然不同，美国政府不对中国提供环境项目的贷款或援助款。由于美国国会限制对中国进行直接援助，因此将近20个美国政府机关目前在中国开展的100多项环境或能源活动都是通过机构内部预算实现，而不是通过官方发展援助实现。尽管资金有限，但根据美国和中国在1979年的《科技合作协定》(1979 U.S.-China Scientific and Technology Cooperative Agreement)，中美签订了30项议定书，作为双方合作项目、研究和信息交流的基础，内容涉及自然资源保护、大气，海洋环境，污染和能源等。虽然空气质量和能源有效利用是合作，研究和交流的重点领域，但在过去几年来，美国的机构也在中国开展了一些水资源的综合项目，以下将对此作概括介绍。

美国农业部与环境保护局

从2000年以来，美国农业部 (USDA)与环保署 (EPA)在黄河下游开展了水质监测，废水再利用和流域管理的示范项目。举例来说，自2001年以来，美国农业部 和环保署 与中国水利部及地方环保局合作，在山东和河南共同推行废水处理与监测的示范项目。

美国农业部的经济研究项目一直与中国科学院，水利部，澳大利亚农业与资源经济局，加州大学戴维斯分校合作，研究中国的水资源与农业生产的课题。从2003年以来，此项合作的重点是为黄河流域模型搜集数据，并在2005年模拟了流域内水权交易和环境流的初步情景。该情景指出水权交易可以促进谷物生产。这些合作伙伴还在流域内调查节水技术的应用，使用者协会的创立以及灌溉区域的运河发包工程的改革。

2006年，美国农业部的经济研究项目与中国水利部共同提出一个新计划，试图理解灌溉管理改革和节水灌溉技术的运用可能产生的水文后果，并将其影响包括在黄河流域模型中的水文部分中。此外，双方还共同研究中国农村实际上存在的水权现象，并建议采用与实际情况紧密结合的正式水权制度。

2006年，美国环保署将在海河流域完成"中国可持续性城市的净水计划"。通过与天津环保局、国家环保总局，水利部、海河保护委员会，全球环境机构以及亚洲开发银行的合作，本计划主要对水质给与关注。本计划旨在扩大安全饮用水和卫生设施的供应，完善海河流域靠近天津地区的流域管理。本

在中国，媒体是公众了解环境问题的重要渠道。很多有责任感的新闻记者也把环境保护作为他们关注和报道的重要内容，在污染和生态破坏的现场，经常可以看到他们的身影。图为一些摄影记者正在中国中部某省份拍摄一条严重污染的河流。图片来源：《环境保护》图片资料

计划还将通过提高对附近村庄，酒店，饭店，鱼塘及农业环境废水和流失物的管理，来专注于提高于桥水库的水源质量。在与全球环境机构海河流域水资源和环境综合管理项目的合作下，本计划将推进流域管理计划的形成。

一项新的水污染防治行动计划诞生于中国环境总局与美国环保署在2003年签署的双边协议。协议包括在中国开展污染交易试点项目的双边备忘录。

在中国从事水资源相关工作的国际环境非政府组织及研究机构

过去几年以来，国际非政府组织开始在中国从事越来越多的流域保护与管理工作。虽然美国的环境非政府组织将中国流域保护作为主要或次要的工作范围 (参见表五)，但日本的环境非政府组织和研究机构则围绕中国流域的大主题(参见表六)，积极组织研究考察团，开办学术会议和研讨会。这些国际非政府组织与研究机构的水资源项目致力于建立网络，以联合中央，省级和地方政府机构，研究中心和中国非政府组织。总体来说，这些项目创造了新的沟通渠道，并扩大了各利益相关方对中国水资源保护工作的参与。

第IV部分　日美在中国流域治理中的合作机会

　　过去十五年来，为了清理整治主要河流及湖泊，中国政府不断推行强有力的水资源保护政策，设定了宏伟的目标及措施。尽管如此，中国的水质，特别是河流的水质，还是在显著恶化。国内水资源法律法规的改革以及国际援助推进了水资源综合管理理念在中国的发展。

　　要确保缓解中国水资源问题，特别是流域保护问题所必需的政策改变和国际援助，首先要鼓励创新性的观念，促进来自国际，区域，国家，地区各组织的环境专家和业内人士的对话。

　　在未来的几十年中，协助中国走上可持续流域开发的道路具有重要的意义。因此日本及美国应在最大程度上寻求合作，或者至少协调各自的活动，来交流彼此的技术及经验。

　　美国和日本政府（同时还有非政府组织和研究机构）一直积极投入关于中国环保（特别是水资源）的协助和研究工作。但两个国家在该领域几乎没有信息共享，也没有正式的合作活动。随着经济发展的减速，政治重点的变迁，以及最近的主要自然灾害，迫使美国和日本消减海外发展援助资金。因此，在国际环保协助工作中，美日信息共享和合作活动可以让正在缩减的援助资金发挥更大的作用，也可以避免在中国及其它发展中国家项目建设的重复投资。

　　下文首先讨论了日本及美国政府在国际援助项目中将水资源放在首要地位的原因。之后讨论了日本和美国政府，民间组织及研究中心在中国河流保护方面可能进行合作的领域。在中国协助建设综合流域治理的美日合作关系可以围绕三个主题开展：（1）流域治理体制，（2）融资机制，（3）公众参与。

水资源问题在国际援助项目中占重要地位

　　美国和日本都将水资源问题放在国际援助项目中十分重要的位置，通常将其作为发展中国家扶贫和城市化发展的一个方面考虑。2003年在日本召开的第三届世界水资源论坛(Third World Water Forum)的一项重要建议为在发展中国家加强水资源问题的国际合作。根据该建议的精神，日本水资源机构和亚洲发展银行在世界水资源论坛增设了亚洲流域组织网络（NARBO）项目。借鉴日本在水资源开发及保护方面的经验，亚洲流域组织网络的目标是通过倡导，培训，技术咨询及区域合作在亚洲的河流流域促进水资源综合管理。

　　美国国际发展协会(USAID)也将世界水资源的保护及合理开发放在首要位置。在世界许多国家，美国国际发展协会在水资源方面的项目和投资主要集中在提高安全充足的水源供给及卫生设备的普及性，提高灌溉技术，保护水体生态系统，加强水资源管理体系的能力。在2003至2005年间，美国国际发展协会在76个发展中国家共投资17亿美元来提高淡水及海岸资源的可持续管理。同一时期，2400万人的淡水供给得到改善，2800万人的卫生系统

得到改善，3400个流域管理组织召开会议并决定采取流域范围的水资源综合管理决策方式。鉴于2005年11月30日参议员保罗·西蒙的《2005年贫困人口水资源法案》已被正式立法，美国国际发展协会在水资源方面的工作很有可能得到扩展。该新法案的目标是将提供安全、价格低廉的饮用水和卫生设备，以及可持续的水资源管理作为美国对外政策的基石。

除了独立的国际援助项目外，美国和日本都在不断寻求新的方式以加强水资源项目的合作关系。2002年的可持续发展世界首脑会议中，美国和日本政府开始了一项新的水资源合作项目（美日水资源合作关系）。两国同意在发展中国家中采取联合或平行的水资源项目开发。美国国际发展协会和日本国际协力银行正在为在四个国家——菲律宾，印度尼西亚，牙买加和印度施行水资源融资项目进行努力。菲律宾已有两个试点项目在进行当中。其中一个项目中，市政水资源贷款融资机构将利用日本国际协力银行支持的信贷设施以及由美国国际发展协会的发展信贷权威机构支持的个人投资。同时，在2005年初，预计在2007年施行的菲律宾水资源循环基金项目完成了可行性研究调查。其它三个试点的类似融资项目也正在设计之中。虽然中国目前不在该合作项目的范围之内，但显而易见，中国能够从美日水资源融资援助项目中受益。

潜在的合作领域

中国对水资源的需求巨大而复杂。但通过流域综合治理的镜头，聚焦于管理体系、融资机制及公众参与，希望本文所讨论的一系列方法能抛砖引玉，刺激美国和日本（及其他国家）对于在中国开展水资源合作项目方面更多好想法的产生。援助合作的潜力不仅存在于美日的政府机构，也存在于两国的民间组织和研究机构中。

立法和体制改革

在利用立法和体制改革以促进流域综合治理观念在中国的普及上，美国和日本政府可以共同努力，聚焦于从七个设有水利委员会的主要流域地区选择一个（例如支流，湖泊和河口）开展试点项目。试点项目可以小规模的体制改革为焦点，例如确立水权，成立用水者联合会和建立水价。中国环境与发展国际合作委员会的流域综合管理任务组甚至建议了一个更为宏伟的试点项目。该任务组提议创立一个支流，湖泊，或河口范围的，包括省级政府，地方当局以及利益相关方代表的的管理委员会。这些地区性流域管理委员会将承担制定流域规划与目标，监督群众参与的试验，以及提供鼓励流域保护的经济刺激等责任。

国际交流项目可以提高流域管理委员会由下至上试点改革的成绩。交流项目可以为来自监督项目实施的流域委员会的人员提供在美日流域管理委员会工作数月的机会。对具备包括流域地区所有利益相关方机制的流域管理委员会的访问不仅可以使中国的河流管理者洞察到如何在工作中实现更广泛的公众参与，也可以让他们学习到如何预防与解决由水资源引起的纠纷。中国在解决国际及国内水资源纠纷中面临的问题越来越尖锐，部分原因是中国政府将注意力狭隘地集中于流域管理的经济效益方面，而不是以发展的方式协调人类与生态系统的需要。作为比较，日本、美国和其他发达国家正在逐渐将重点转移到河流流量的生态价值上。美国和日本也要面对流域纠纷问题，但两个国家都已建立起法律体系强调河流流量的价值，也

为群众参与和纠纷解决提供了正式的渠道。

尽管与中国的七个主要流域相比，日美的流域面积要小得多。但是我们仍然相信，日美的流域管理经验在中国的亚流域地区是适用的。值得一提的还有中国的流域管理委员会自20世纪50年代成立以来实行的改革也能对美国和日本制定国内和对外援助政策提供一些经验。特拉华流域委员会是一个特别值得学习的美国流域委员会。自1961年成立以来，特拉华流域委员会及其成员（四个流经州及联邦政府）不仅解决了各州间的纠纷，还是动员政府，公民及各民间组织解决水资源短缺及污染问题的有效渠道。与中国各流域委员会不吸收省份作为成员，缺少足够的权威与广泛性等特征相比，特拉华流域委员会展示了一个有趣的管理模式，即如何利用充分的管理权威和利益相关方的参与来实现流域的优化管理。

中日两国都实行中央集权的河流管理体系，因此日本的流域委员会应能为中国流域管理提供宝贵的经验。自1997年新的《河川法》要求成立流域委员会以来，许多湖泊和河流地区都成立了自己的流域委员会。虽然这些流域委员会都是新近成立的机构，但是它们在就流域开发的敏感问题和流域环境保护组织利益相关方共同达成共识方面已经具备了相当的经验。例如淀川流域委员会就是一个独特的咨询机构。自成立以来的四年半间，一共组织了400多个对公众开放的流域管理计划会议。虽然淀川流域委员会是由土地与交通部管辖的地区发展署建立的，但它并不由该署管理，而是通过与包括学者、社区及民间团体代表的成员组织的共同协商进行管理。该委员会的行政工作是由一个私有公司负责。公开的群众导论方式减慢了淀川流域委员会的工作进度。委员会目前仍在拟定流域管理计划的草案。但一旦得以实施，该计划几乎不会遇到阻力，还将简化争端的解决。

我们同意中国环境与发展国际合作委员会（CCICED）的流域综合管理任务组报告中提出的一项建议，即除了实行区域试点以外，更多的国际援助可以聚焦于全国范围内水资源管理的体制和法律改革。例如，国际援助方可以和中央政府建立伙伴关系，帮助阅览与修改关于流域管理及水污染控制的立法，以减少机构矛盾，并明确各流域委员会的相关责任。法律修改的一种合作方式可为部门间的互动。例如，人民代表大会环境保护与自然资源委员为的成员可以与美国与日本的同僚会面，以了解两国在河流及水资源保护上采用的有效的法律体系（如美国的《河流自然景观法案》）。

流域综合管理任务组提出的另外一项对付"多头龙"的潜在的举措就是成立一个国家级的流域综合管理委员会。该委员会将包括国家发改委，国家水利部以及国家环保局。委员会将监督法律的修改与制定，促进在全国范围内实现流域综合管理。

正如在其它国家一样，危及的出现——比如松花江的剧毒污染事件，以及黄河的几次主要断流——都可以团结政府各部门开展河流保护工作。可是，在中国为解决河流危机所产生的合作往往会加剧河流管理的中央集权化，导致低效率的措施。这些都不利于建立可持续水资源治理的综合体制。为政府各部门间的合作创造动力是在中国建立强有力的流域综合管理体系的关键。为促进各部门的合作，必须在对以下两方面进行研究：（1）首先，确定各职能部门具体在何时，以何种方式合作能提高社会效益；（2）怎样有效的动员政治力量，促使各职能部门在必要的时候开展合作。中国应迈出的最切合实际的第一步就是信息和数据的共享。这不仅有助于水资源保护政策的拟定和实施，也节约了解决水资源问题的资金和时间。因此最终对所有利益相关方都是有利的。

利用新的融资机制与动力

"谁受益，谁出资"是讨论可持续流域管理的经济学原理的重要问题。为了回答这一问题，中国的政策制定者们热心于引进市场机制作为促进环境法规执行和环境保护（特别是水资源保护）的新工具。作为促进水资源保护的方法，水权的交易吸引了中国许多技术人员和学者的注意力。但是，中国水权的不明确和法律体系的薄弱阻碍了系统化的水交易市场的形成。

中国的水权是一个复杂而敏感的话题。美国农业部和日本国际协力机构目前都有独立项目，旨在明确中国的水权。因此，即使正式的合作项目不具备可行性，美国农业部及日本国际协力机构在该领域进行试点项目和研究考察的结果应被更广泛的分享和传播。信息共享有助于确定在该领域可能的合作项目。

为了收回建造和运行供水系统和污水处理场的成本，中国应更广泛的制定水价。目前在水资源保护方面，中国似乎还没有较好的成本分摊模式。这是目前中国可持续流域管理所面临的迫切需要解决的难题。日本及美国的政府和非政府组织可与中国政府合作，建立促进河流及湿地资源保护的资金支持机制。虽然中国政府对此表示出极大的兴趣，但目前几乎没有任何国际或国内的行动，来促进水资源保护的融资机制（例如环境服务偿付机制，绿色税收，循环基金及市政债券）和市场机制（例如水权交易和水资源储存）的形成。在面临这些水资源融资的难题下，美国，日本，中国在若干潜在领域的合作机会阐述如下。

建立环境服务偿付试点。怎样建立有效的机制，刺激下游用水方对上游开展的水土保持工作进行补偿是许多国家共同面临的挑战。美国可以提供这方面的成功事例。美国国际发展协会项目使用的模式也支持了无数发展中国家的环境服务偿付试点项目。另外一种模式为旨在提高上游水资源和森林保护的税收体制。这种模式被很多日本都道府县级政府所采用。由于这样的"绿色"税收机制在日本还只实行了两年，因此对其就河流保护有效性进行评估还为时过早。但是中国的政策制定者仍能从这些机制中学到，怎样以上下游利益相关方的合作为基础，引进经济刺激模式来保护水资源。类似的机制在中国其实已有先例。中国政府已经实施了一些赔偿政策（主要为收费，国家补助，税收及惩罚制度）来保护森林资源。例如，伐木在中国西南的大部分地区已被禁止。政府还通过对农民发放额外的谷物和补助鼓励退耕还林。但是，这些森林保护项目都为政府项目，而非市场主导项目。

扩展关注水量和水质的用户联盟。虽然世界银行帮助在中国建立了近2000个水资源使用者联盟，但农村地区的许多地方水资源管理局在提供足够的服务和评估水资源使用费中还是面临严重的困难。因此显而易见，有必要对国外其它成功的水资源使用联盟模式，特别是促进污染控制的模式，进行考察。例如在荷兰，水资源委员会组织由地方各利益相关方构成，并在制定排污收费，分担投资方面起着重要作用。

建立循环基金支持水资源保护和污染控制活动。1987年，美国国会修改《净水法案》，一个具有创新性的州立净水循环基金（Clean Water State Revolving Fund）项目也应运而生。州立净水循环基金对很多水质项目提供资金支持，其中包括无点源污染，流域保护及还原，河口管理项目，以及较为传统的市政污水处理项目。成立类似基金是中国一些试点项目的目

标。例如2004年，作为长江中游湿地恢复工作的一部分，世界野生动物基金会中国分会建立了一支循环基金，旨在帮助在湿地恢复工作中失去土地的秋湖村村民开发其它生计。基金的本金帮助了当地农民开发竹园，可持续渔业，生态旅游以及水耕蔬菜种植。在第一轮贷款周期中，104户农户还清了贷款。偿付的利息被积累起来，以便满足其他农户的借贷需求。

发行市政债券为污水处理场集资。在控制水污染方面，中央政府缺乏帮助污水处理场收回部分成本的正式政策。污水处理被看作地方经济发展的阻碍，因此地方政府也不愿意对污水处理进行投资。一项潜在的解决方法就是发行市政债券为污水处理场等环保基础设施集资。这要求中国在法律和融资体系方面作出较为重大的改变。世界银行和美国贸易发展署在上海完成了一个试点项目，成功发行了市政债券为污水处理集资。类似的举措也可以在其它城市开展。为了便于对现行的财政税收法律进行调整，以鼓励投资者对市政债券市场的参与，并创造有效的市场和机制来减少市政债券的投资风险，市政债券项目应寻求与市政府与国家发改委的合作。

在小范围内尝试水权交易。正如在第II部分中所述，在中国一些地方没有法律保护的水权交易正在悄悄进行。美国与日本可以利用目前在水权交易方面的经验，帮助在地方范围内建立水权交易的基础体系。

在环境影响评估和计划中更广泛的反映投资与收益。根据中国环境与发展国际合作委员会的流域综合管理任务组的建议，国际上的融资行动可以专注于某一流域。最理想的情况是，流域委员会的开发与规划计划以环境影响评估为基础，不但要重视经济成本，也要重视社会及环境的成本和收益。美，日，中可以合作开展调查研究和试点项目，以辨认在河流管理中包括环境，生态及社会影响评估会遇到怎样的障碍及潜在的解决方法。

开辟更广泛的群众参与平台

从90年代中期以来，中国政府不断积极鼓励环境领域的公众参与。这不仅是因为许多国际项目促进了政府和公民间的合作，也因为改革时期发生的政治变化创造了更开放的社会风气。有多项改革刺激了环保工作的公众参与，并显著改变了群众与国家的关系，其中包括：更开放的新闻媒体，允许个人成立民间团体的规定，发生个人伤害时的起诉权，环境影响评估必须包括群众意见，以及逐渐开放的信息渠道。这些趋势都为致力于普及流域管理公共参与的国际合作创造了机会。第II,III部分给出了国际国内的民间组织及双边援助工作在推进公共参与方面的实例。同时中国环保总局和其它部门也承认，更多类似的工作应该得以开展。群众参与流域管理的基本要求是流域的所有利益相关方都有开放的信息渠道，并且在流域综合管理规划，环境影响评估及流域管理决策中享有发言权。美日政府部门，民间团体以及研究机构可以找到很多机会来协助在水资源管理及污染控制中推进更广泛的群众参与。

（1）流域范围论坛。中国环境与发展国际合作委员会的流域综合管理任务组建议在每个面积较大的流域都设立发展及保护论坛，目的是为不同的流域省份、政府部门、民间团体及研究机构提供交流和协商的平台。

（2）促进企业社会责任在亚流域或市政范围的论坛。工业企业往往是主要的排污者。要动员它们或地方政府与社区团体、民间组织或高校合作，采取防治措施又十分困难。不过国际组织可以在亚流域或城市范围内开展活动，教育工商业和其它利益相关方，让它们意识到保护水资源的措施最后

是可以为企业带来经济效益的。企业社会责任在污染控制方面具体表现的实例包括：自愿超额完成减少排污的指标，将排污信息向大众公开，采用绿色供货渠道，建立民间团体与工业企业间的合作关系，自愿参与污水交易的试点项目，采取透明的应急管理措施。

（3）流域间的交流。负责流域管理的官员和民间组织可以参加赴美国或日本参观考察的交流项目，学习如何在河流和流域管理中加入公共参与。

（4）召开听证会以确定流域管理决策。在流域综合管理试点项目中，关键是要建立公共参与的机制，定期参与流域管理方案的制定与施行。目前的听证会都是在政策讨论结束以后举行，主要让群众发表评论。2005年11月，中国环保总局在国际范围内寻求法规设计的建议，以便在环境影响评估中扩大公众参与力度。一旦新的法规出台，环保总局将寻求协助以便开展培训。

（5）对污染受害者的法律援助。在水污染事故发生后帮助公民走向法庭能够给地方政府和工业企业施加压力，迫使它们执行现有的水污染保护法。目前为止，中国只有唯一的一个民间组织在致力于为污染受害者提供法律援助，因此法律渠道在中国的运用还是鲜为人知的。

（6）培训中国的民间组织。中国环保性民间组织在能力和效率上都有了很大的提高，这在很大程度上要归功于国际民间组织，基金会和政府部门的支持。国外援助十分宝贵，因此更多专注于水方面的工作应该开展起来。比如，目前就有很多双边或多边的流域保护活动正在进行，但是极少数有民间组织的参与。带动民间组织的参与对于确立它们在流域管理中的合法地位至关重要。

(7)培养主人翁意识。促进政府及公众的合作，让他们共同对流域进行管理固然重要，但赋予公民更大的权利，让他们成为河流和近水陆地的保护者也具有重要的意义。越多的公民对流域资源表示出关心，政府对河流保护投入的成本就越低。可以借鉴的经验有 日本拉姆萨中心(Ramsar Center Japan)在日本及印度河口地区开展的工作。拉姆萨中心。不仅仅促进了政府和公众间成功的伙伴关系，并且赋予公众实际的权利，让他们成为怎样恢复近岸珊瑚岛生态系统并同时提高生活水平的主要决策者。在水资源管理中发挥主人翁意识是对以管理为主的手段的明智补充，代表了实现长期目标的希望，也是长期保持流域范围内基本的生态产品和劳务所需的不间断努力的保障。

结论

作为一个强有力的新兴经济和"世界工厂"，中国对全球市场的影响——在进出口两方面——都是巨大的。中国的影响在以后十年还将继续扩大。中国政府今天作出的环境保护和能源维护的决定将在将来产生全球性的影响。中国在环境和能源方面的立法已经取得了长足的进步，足以将中国变为可持续发展的典范。不幸的是，这些法律的实施极端不平衡，特别是在水资源方面。

中国面临的水危机挑战是严峻的。中国政府对水资源管理内部改革的开放态度及对国外缓解水资源问题模式的不断探索为日本，美国及其他国家协助中国开放了渠道。美日可以为中国提供流域管理不同的背景和经验，激发更好的保护中国河流的想法和方案。总而言之，我们相信日美为促进可持续流域管理跨太平洋的合作不仅能为中国的水资源安全做出贡献，也能为世界范围的水资源安全做出贡献。